WEYERHAEUSER ENVIRONMENTAL BOOKS

Paul S. Sutter, Editor

WEYERHAEUSER ENVIRONMENTAL BOOKS explore human relationships with natural environments in all their variety and complexity. They seek to cast new light on the ways that natural systems affect human communities, the ways that people affect the environments of which they are a part, and the ways that different cultural conceptions of nature profoundly shape our sense of the world around us. A complete list of the books in the series appears at the end of this book.

COMMUNIST PIGS

An Animal History of East Germany's Rise and Fall

THOMAS FLEISCHMAN

UNIVERSITY OF WASHINGTON PRESS
Seattle

Communist Pigs is published with the assistance of a grant from the Weyerhaeuser Environmental Books Endowment, established by the Weyerhaeuser Company Foundation, members of the Weyerhaeuser family, and Janet and Jack Creighton.

Copyright © 2020 by the University of Washington Press

Composed in Minion Pro, typeface designed by Robert Slimbach
Cover design by Amanda Weiss
Cover illustration by Amanda Weiss based on illustrations from *Cyclopedia of Farm Animals* by L. H. Bailey (New York: Macmillan, 1922) and on clip art courtesy FCIT (bottom pig).
Map of the German Democratic Republic courtesy of Meghan Cohorst

24 23 22 21 20 5 4 3 2 1

Printed and bound in the United States of America

All rights reserved. No part of this publication may be reproduced or transmitted in any form or by any means, electronic or mechanical, including photocopy, recording, or any information storage or retrieval system, without permission in writing from the publisher.

UNIVERSITY OF WASHINGTON PRESS
uwapress.uw.edu

LIBRARY OF CONGRESS CATALOGING-IN-PUBLICATION DATA
Names: Fleischman, Thomas, author.
Title: Communist pigs : an animal history of East Germany's rise and fall / Thomas Fleischman.
Description: Seattle : University of Washington Press, [2020] | Series: Weyerhaeuser environmental books | Includes bibliographical references and index.
Identifiers: LCCN 2019045350 (print) | LCCN 2019045351 (ebook) | ISBN 9780295747309 (hardcover) | ISBN 9780295747316 (ebook)
Subjects: LCSH: Agriculture and state—Germany (East) | Agriculture and politics—Germany (East) | Communism and agriculture—Germany (East) | Factory farms—Germany (East) | Swine—Breeding—Germany (East) | Pork industry and trade—Germany (East)
Classification: LCC HD1960.5 .F54 2020 (print) | LCC HD1960.5 (ebook) | DDC 338.1/76400943109045—dc23
LC record available at https://lccn.loc.gov/2019045350
LC ebook record available at https://lccn.loc.gov/2019045351

The paper used in this publication is acid free and meets the minimum requirements of American National Standard for Information Sciences—Permanence of Paper for Printed Library Materials, ANSI Z39.48–1984.∞

For Iris, my star witness and constant companion

CONTENTS

FOREWORD

All Pigs Are Ideological, but Some Pigs Are More Ideological than Others

PAUL S. SUTTER

Donald Worster began *Dust Bowl* (1979), his landmark history of the ecological destructiveness of capitalist agriculture on the American Great Plains, with a quote from Karl Marx. "All progress in capitalist agriculture," his opening epigraph read, "is a progress in the art, not only of robbing the laborer, but of robbing the soil." With great economy, Marx had summarized one of Worster's central arguments: that by conceptualizing nature merely as capital-in-the-making, capitalists exploited the land as well as labor. For Worster, the dust storms of the dirty thirties were not a natural disaster, the products only of drought, heat, and wind; they were the returns on an economic system that had ignored and thus undermined the natural order on the southern plains in pursuit of wealth. Indeed, Worster's unsparing critique of a "culture of capitalism" was of a piece with a major message of the first generation of environmental history scholarship: that the commodification of nature under capitalism had been the major driver of modern environmental degradation.

Dust Bowl was a staple in my undergraduate and graduate courses during my early career, and as I walked through the book's arguments with my students—most of whom had come of age in the immediate aftermath of the fall of the Berlin Wall and the collapse of the Soviet Union—I came to anticipate an invariable question: "What about communism?" For a few students, the question drew on a chauvinistic unease with an interpretation that so relentlessly critiqued America's prized, and apparently victorious, economic system. But for others, most of whom were willing to accept the argument that capitalism was environmentally destructive, the question evoked a manifest impression of the post-Soviet era: that the environmental degradation of the Western world paled in comparison to the dreary wreckage of the communist East. While I tried to explain that the Soviet Union and other Eastern Bloc countries had command economies that similarly commodified nature, and that we ought not to let Cold War binaries get in the way of seeing that commonality, that was a difficult point to make, precisely because the Western world's Cold War triumph seemed as environmental as it was economic and ideological. When Americans thought of the failures of the communist system, they were as likely to summon images of air grossly polluted by the combustion of brown coal as they were of political repression and barren grocery store shelves. Such grim assessments particularly attached themselves to the former East Germany, whose greener neighbor to the west made it seem the apotheosis of communism's environmental destructiveness.

In *Communist Pigs*, his fascinating history of how pigs were central to the rise and fall of the German Democratic Republic (GDR), Thomas Fleischman demonstrates that a market-oriented version of industrial animal production had won a Cold War victory in East Germany well before the fall of the Berlin Wall. Indeed, in an opening vignette that makes brilliant use of George Orwell's *Animal Farm* as a kind of prophecy, Fleischman suggests that industrial pig production in postwar East Germany came to stand on the same hind legs as capitalist pig production in the West. On one level, this is not surprising. *Communist Pigs* is a case study in a now well-developed literature on how the former Soviet Union and its satellites came to embrace an industrial ideal in agriculture pioneered in the United States. For Fleischman, when it came to the rise of industrial pork production, East Germany and Iowa were, to quote Kate Brown, "nearly the same place." In terms of their methods and their results, communist and capitalist pig farming were not only all but indistinguishable but also similarly entangled in global export markets. They were competing systems that were also, ironically, parts of the same system. While *Communist Pigs* appears as the third installment in an accidental trilogy about the ideological nature of pigs under different modern political systems—it joins Tiago Saraiva's *Fascist Pigs* and J. L. Anderson's

Capitalist Pigs among recently published books on the subject—Fleischman also insists that we not give ideology too determinative a role. His is a tale of an East German industrial food system that, beneath the veneer of Cold War ideological divisions, shared a deep kinship with the West. More than that, it is a story in which pigs transformed, and in some cases undermined, the meanings and practices of communism. A history of communism with pigs at its center is not so tidy.

Fleischman's communist pigs are a diverse and contingent bunch. While the food system of the GDR was built on a modernist desire to master the efficient production of porcine flesh, it also opened up spaces for two other kinds of pigs to thrive in postwar East Germany. First, there were garden pigs, raised on the traditional small plots of East German farmers to augment their austere lives. While the leaders of the GDR feared that such private property in animals was a dangerous remnant of bourgeois economic relations, they tolerated these animals as a necessary evil as they redirected industrial pork production toward the export market. As the GDR traded their industrial pork for the capital that they needed to undertake other projects, garden pigs became essential tools for self-provisioning, and their East German owners came to see them as a customary right. At the same time, East Germany's wild boar population skyrocketed as concentrated and confined meat production transformed agricultural landscapes. These other two pigs, Fleischman suggests, were certainly produced within the communist system. But garden pigs and wild boars—physiological variants on the same species that industrial producers were bending to suit their system—enter this history as the industrial pig's alter egos, as symbols of the system's leakages and shortcomings. Indeed, the garden pig became a focal point in discussions about the nature of communism and a vehicle for East Germans to make demands of the regime even as the wild boar raised profound questions about who and what belonged in communist nature.

If one of Fleischman's signature arguments is that industrial animal agriculture in East Germany was all but indistinguishable from that of the West, another is that East German pork production was captive to the vicissitudes of the international markets in which it became entangled. This is a complex story, but a vital one, and Fleischman tells it well. Industrial pork production grew slowly as a borrowing from the West in the quarter century after World War II, but it climaxed in the 1970s as a specific set of market conditions—including extraordinarily favorable credit and access to cheap Western grain—came to reshape East German agriculture. Up to that point, concentrated pork production had been limited by the capacity of East Germans to raise the necessary fodder on a limited land base, but Western grain briefly eliminated that constraint and allowed industrial pork production to expand dramatically. Such an expansion, however, only worked as long as the spigots of cheap grain and ready capital remained open. Once they closed, the system began to implode, and so did the nation. In the end, it was a system built of sticks and straw.

"There was no place in the country untouched by pigs," Fleischman notes of postwar East Germany. This was certainly true for an economy that made industrial pork production central to its agricultural profile, and for the garden pigs and wild boars, which also proliferated as a result of this system. But it was most tragically true of the pollution produced by this system. Pig wastes literally swamped the nation, their sheer volume outstripping the absorptive capacity of land and water. Marx called this problem "metabolic rift," a process essential to capitalism whereby a by-product such as manure, which might have restored soil fertility in a virtuous closed loop, gets disconnected from the healthy metabolism of the land and becomes waste. To put it bluntly, East German pig production became the posterchild of metabolic rift. As nitrate pollution increased, so did nitrate poisoning, while massive piles of pig manure—"mossy giants" as they came to be known—became ubiquitous topographic features throughout the landscape. By the late 1980s, East Germans had had enough, and they began to organize an environmental opposition that was critical to the undoing of the communist regime. As Fleischman shows in remarkable detail, pigs of various sorts made the East German economy under communism, and they would help to unmake it as well.

So what about communism? How are we to understand its environmental impacts as compared to those of capitalism? As an East German case study, *Communist Pigs* provides several answers to this question. One answer is that communism was often capitalism by another name—"state capitalism" is what Fleischman calls it—and that the fall of the GDR was less an object lesson in the superiority of capitalism than it was a harbinger of capitalist environmental stresses to come. As such, narratives of communism's environmental destructiveness and capitalism's green Cold War victory are opiates numbing us to the reality of a continuing crisis. As Fleischman reveals at the end of *Communist Pigs*, while the GDR is no more, its system of industrial pork production is alive and well in a reunified Germany—and in many other parts of the world as well. But perhaps the more innovative and satisfying answers Fleischman provides are ones that push beyond the ideological categories at play and highlight the contingencies that explain East Germany's postwar environmental history. The German Democratic Republic fell apart not for ideological but for historical reasons, as a result of specific choices by East Germans and of agencies and forces beyond their control. While George Orwell may have been prescient in foreseeing the ideological drift of communist pigs toward capitalist modes of production, East Germany's pigs also acted beyond ideology, and so, often, did East Germans.

ACKNOWLEDGMENTS

Even as I type these words, it is difficult to wrap my head around the idea that this book is finally done. Ten years is a long time by any measure, and it is an especially long time to spend working on a single book. I am realizing, however, that I needed every one of those years so that I could meet the many people whose views, suggestions, and ideas have so profoundly influenced this work. I am most grateful for the guidance of Mary Nolan, who has shaped my professional life and showed me that it's possible to be a great scholar, courageous activist, and supportive mentor all at once. Karl Appuhn introduced me to the field of environmental history. In addition to being a faithful reader and editor of my work, he acted as a relentless booster on my behalf, introducing me to his peers and telling anyone who would listen about my work. Yanni Kotsonis established my baseline of historical theory, shaping how I think as a historian. He has also broken bread with me on multiple occasions to offer a free meal and advice to a poor graduate student. And when I needed refuge after a year of striking out on the job market, he found me a professional title, an office, and library privileges to tide me over for another year.

This project would not have been possible without my editorial team at the University of Washington Press. Series editor Paul Sutter saw promise in my original proposal and took more time than is reasonable for a busy academic to exchange emails, send drafts, and talk on the phone, all in the service of getting me to write the best possible version of this book. The care and attention Paul and his whole team give to first-time authors is what makes the Weyerhaeuser Series a truly exceptional place in the competitive field of academic publishing. At Washington, I'd also like to thank Andrew Berzanskis, Regan Huff, and Catherine Cocks for escorting this project over five long years of revision. Thanks as well to the two anonymous reviewers who provided constructive feedback, and Meghan Cohorst for making the map of the GDR.

Several foundations and grants supported the research and writing of this book, including the German Academic Exchange Service, the American Council of Learned Societies, the Mellon Foundation, and the Global Research Initiatives Fund at New York University. These programs funded multiple trips to Washington, DC, and Berlin, Germany. Thank you to the archival staff at the National Agricultural Library, the Library of Congress, the Bundesarchiv and Bibliothek in Berlin, as well as the Brandenburg/Berlin Landesarchiv. I would also be remiss if I did not thank the innumerable writing spaces, from coffee shops to libraries, that broke up the monotony of working from home. Thank you especially to Root Hill Café, Konditori, and Vbar in Manhattan and Brooklyn, Equal Grounds Café in Rochester, New York, the Emporium in Yellow Springs, Ohio, and the Mercantile Library in Cincinnati and the Guilford Public Library in Connecticut.

Several writing workshops provided a venue to share my work. The now tragically defunct Dissertation Proposal Development Fellowship with the Social Science Research Council helped get this project off the ground, introducing me to people from a variety of academic disciplines with whom I continued to work long after the fellowship was over. Thank you especially to Jennifer Baka, Gabrielle Clark, Aaron Jakes, Greta Marchesi, W. Thomas Okie, and Sara Safransky. I had the honor of spending the 2013–14 academic year with the Program in Agrarian Studies at Yale University, attending the Friday morning colloquium in New Haven and getting to know brilliant people, including Jessica Cattelino, Federico D'Onofrio, Alan Mikhail, Abby Neeley, Andrew Offenburger, James Scott, Kalyanakrishnan Sivaramakrishnan, and Jinba Tenzin. At Yale, the Center for Earth Observation advised me in the creation of the remote sensing maps in this book. Thank you to Larry Bonneau and assistant Frieda Fein for their patience and hard work.

I am grateful for the many friendships I made during my short time at Bowdoin College that endure to the present. Thank you to Todd Berzon, Connie Chiang, Shaun Golding, Cory Gooding, Tristan Grunow, Brian Kim, Jens Klenner, Barbara Elias, Matt Klingle, and Kathryn Sederberg. I'd also like to thank my New York City colleagues, Natalie Blum-Ross, James

Cantres, Sasha Disko, Liz Fink, Laura Honsberger, Devin Jacob, Masha Kirasirova, Anna Koch, Nathan Marcus, Shaul Mittelpunkt, Kate Mulry, Eric Owens, Jess Pearson, David Rainbow, Samantha Seeley, Anelise Shrout, Quinn Slobodian, Evan Spritzer, Geoff Traugh, Matt Watkins, Jerusha Westbury, and Peter Wirzbicki. In New York and beyond, David Huyssen, Scott Moranda, Catherine McNeur, Eli Rubin, Neil Maher, Atina Grossman, and William Germano offered guidance and friendship in the early stages of my career.

In 2016, I had the great good fortune to be hired by the Department of History at the University of Rochester. Thank you to Joan Rubin, director of the Humanities Center, for organizing a book manuscript workshop, and to Kate Brown, Donna Harsch, and Andrew Zimmerman, who generously agreed to read the manuscript and then travel to Rochester to give their feedback in person. I am fortunate to have found such kind, supportive colleagues in the Department of History who have made me feel at home. Thanks to my chairs Matt Lenoe, Stewart Weaver, and Laura Smoller, as well as other colleagues from Rochester: Cary Adams, Molly Ball, Camden Burd, Leila Nadir, Steve Roessner, Pablo Sierra Silva, Blair Tinker, and Brianna Theobald.

This project could not have happened without the support of many people in Germany. Thank you to Heinrich Sevecke in Karcheez, Malte Groth and Hartmut Wenger in Lübz, who gave me tours of their former collective farms. Thomas Paulke of the Deutsches Schweinemuseum sent me several books and images related to the history of pigs. Gerd Lutze, who worked in Eberswalde and now writes local and regional environmental history, shared his expertise, experience, and photographs of the SZMK Eberswalde. Thank you to Hans-Jürgen Rusczyk and other members of the Heinz Meynhardt Freundeskreis for sharing their photographs and personal memories of East Germany's greatest naturalist. In Güstrow, I am eternally grateful to Sigrid and Wilfried Büchner, who hosted me during my Fulbright year in 2005. In Berlin and Rostock, I made lifelong friends in Katja Mielke, Anke Radenacker, and Matthias Braier. Over the years, I have traveled to many parts of the former East with my friends, each of whom has shared their family histories and memories of growing up in the GDR.

Lastly, I would like to thank my family. It is absolutely true that I would not have finished the final draft of this book without the help of my parents, Mary Fleischman and John Fleischman, each of whom read, critiqued, and copyedited every line. This book would be much longer, more boring, and messier without their help. Thanks as well to my father-in-law, George Bieri, who offered his reading services, and my mother-in-law, Abby Cobb, who has always been game to talk pigs with me.

To Iris and Arlo, the sole reason I do anything at all in this world, thank you. Arlo, who has upended my life in all the glorious ways children do, kept me on task with his constant refrain of, "When will you be done with your pig book, Dad?" I am relieved to finally say, "It's finished." This book, however, is for Iris, who has been on this journey since the start. She has had to sacrifice the most for my career, spending months apart during research trips and teaching jobs, taking care of our old sick dog, and uprooting her life on multiple occasions. What's more, she has done so gladly and willingly each time, making a new home wherever we have landed. She reminds me that work is not me; that it is important to get out of my head and out into nature; to pause and take stock. She is the loveliest thing in my life. I owe her everything I have and everything I am.

Map courtesy of Meghan Cohorst.

COMMUNIST PIGS

INTRODUCTION

Animal Farms

"IT IS A SORT OF FAIRY STORY," GEORGE ORWELL WROTE TO HIS literary agent in 1944, "really a fable with a political meaning."[1] The fable was *Animal Farm*, and in it, Orwell set out to destroy the "Soviet myth" and the cult of Stalin.[2] In our time, *Animal Farm* has become a standard part of any good education. Yet in August 1945, when it was published in the United Kingdom, and in April 1946, when it came out in the United States, it was political dynamite. *Animal Farm* drew praise for its allusions to the history of the Russian Revolution, including livestock portrayed as proletariat rebels, animalism as a stand-in for Marxism-Leninism, and the sketches of Lenin, Stalin, and Trotsky in the pigs Old Major, Napoleon, and Snowball. Most memorable of all was the pigs' ultimate betrayal of the revolution, symbolized in the execution of the exploited draft-horse Boxer. Reviewing the book in the *New York Times* in 1946, the presidential historian Arthur Schlesinger Jr. wrote that it's "a simple story perhaps, but a story of deadly simplicity."[3]

While Orwell's allegory began as a simple story, it did not stay that way for long. For the next half century, politicians and writers coopted Orwell's message to justify their own vision of the ideological clash at the heart of the Cold War. Liberals praised *Animal Farm* and Orwell's later dystopian masterpiece, *1984*, for breaking the European Left's love affair with communism. By the early 1950s, *Animal Farm* and *1984* were being used by US secretary of state Dean Acheson as cultural weapons in the early Cold War. The US State Department disseminated copies all over the world, and Acheson declared that "*Animal Farm* and *1984* have been of great value to the US in its psychological offensive against Communism."[4] In the hands of Cold

Warriors, Orwell's works told of a world starkly divided between dictatorships and democracies, planned economies and free markets, communists and capitalists.[5] Reflecting in 1986 on Orwell's legacy, the American neoconservative Norman Podhoretz declared him "the patron saint of anti-Communism."[6] This would have been news to Orwell, who remained an unrepentant socialist until his death, despite his antipathy for Stalin. To read *Animal Farm* as simple anti-communist satire is to miss the sting at the end. Orwell's allegory is not about a world riven by ideological division. It instead warns about a world dominated by a new permanent ruling class, pigs and men together. The masters of the Orwellian nightmare believed in a hegemonic ideology that fused the authoritarian impulses of Stalinism with the iron hand of capitalistic coercion and exploitation. For Orwell, ideology was not something that existed solely in the communist East and was absent in the capitalist West. It ruled both.

While this theme runs throughout Orwell's works, it is most evident in the other central drama of *Animal Farm*, agriculture. The pig Napoleon's grab for power transforms a small traditional farm into an export-oriented factory. Year after year, the farm grows in size, invests in expensive technologies, and mandates a strict division of labor between managers, technicians, and workers. By the end of the allegory, modernization has made friends out of former enemies. The famous last scene shows the "communist" pigs and the "capitalist" men celebrating together in a drunken rout, with the proletarian worker animals peeking through the window. "The creatures outside looked from pig to man, and man to pig, and from pig to man again; but already it was impossible to say which was which."[7] The animals' observation anticipates a central thesis of this book, *Communist Pigs*, that agriculture under communism came to be indistinguishable from capitalist agriculture, and that pigs provide a clear case study of this convergence. Just thirty years after Orwell's death in 1950, communist agriculture had been drawn into global markets by the fervent adoption of factory farming. From this perspective, one could not tell the communist pigs from the capitalist pigs in the world of industrial pork and agricultural trade in the 1980s.

A second premise of this book is that there is no better place to see animal farm become factory farm than in the German Democratic Republic (GDR). If the word *Orwellian* is shorthand for a kind of totalitarian regime, then the GDR seemed to be its real-world embodiment. The GDR was also a creation of Joseph Stalin. Founded in 1949 on the ruins of the Nazi state, the GDR clearly had Orwellian aspects. There were militarized borders, the ubiquitous secret police, a top-down planned economy, and propaganda-stoked hatred of its Western twin, the Federal Republic of Germany (FRG). In popular memory, the now extinct GDR is remembered for its supposed economic backwardness and the absurdities of its daily life. The East German economy produced horrific pollution, periodic shortages of food and consumer goods, and myriad unwanted or inferior commodities, the plastic-bodied, two-stroke engine Trabant automobile being the most infamous example.[8] In the 1990s, the lack of blue jeans and bananas in the stores became an instant shorthand explanation for collapse.[9] Many contemporaries saw the GDR as a strange, reverse-engineered, mirror-image of West Germany—a real-life *Animal Farm*, fundamentally flawed from the outset and, in retrospect, doomed to failure.

Bananas and blue jeans may have reminded East Germans of the shortcomings of their government, but in 1989, it was pollution that sent them into the streets. International observers and East German citizens cited the GDR's poor environmental record, in particular air and water pollution, as a major source of the government's delegitimization. Many of the grassroots protest groups that began organizing in the late 1970s did so in direct response to this crisis.[10] These organizations later became the springboard for the broad-based popular protest movement that brought down the regime in 1989. Power plants, factories, and factory farms that were poisoning the air, soil, and water were under attack. In 1990 the *Hamburger Abendblatt* reported that "the GDR applied twice as much pesticide and herbicide and three times as much calcium fertilizer per hectare than in the Federal Republic. The four largest hog factories produced more than 4 billion liters of manure annually, the disposal of which is still catastrophic. Hundreds of hectares of forest, fields, and lakes are destroyed."[11] Similar analysis led the national magazine *Der Spiegel* to deride the East German state in its entirety as an "ecological outlaw of the first rank." Discussing the prospects for German reunification, the *New York Times* saw pollution as the fault line dividing East and West, warning that "one issue taking on urgency is how the orderly and clean half of the country can help clean up the disheveled and polluted half."[12]

Since 1990, this view has only become more entrenched, reducing the forty-year history of East Germany into a tale of environmental, ideological, and economic disaster. The dictatorship, by following the distorted imperatives of the planned economy and communist ideology, undermined the freedom and prosperity of its people to produce a one-of-a-kind ecological disaster. As historian Arvid Nelson writes, "Anyone traveling in East

Germany in the 1970s and 1980s could have read the instability of the Marxist-Leninist Ponzi scheme in the Party's cruel and arbitrary human rights policies and in the waste in the country's economy. Pollution was inherent to a communist economy as an unmistakable visible sign of its inevitable decline."[13] The GDR's collapse, like the pig Napoleon's subversion of the animal revolution, stemmed from the rotten ideology and political economy of communism. Even more problematically, it implies that since the collapse of communism, this type of ecological disaster has largely passed from the world—dumped into that "dustbin of history." This perspective teaches us the wrong lessons.

We are not yet free of East Germany's ecological legacy, which tells us about the recent past and the warming world to come. We must reexamine its agricultural and environmental history to understand the origins of the multiple pollution crises that destroyed the regime and persist to this day. *Communist Pigs* locates these crises in the global rise of industrial agriculture in the twentieth century. Like Orwell's pigs and people, communist and capitalist countries adopted analogous forms of agricultural production, which treated land, labor, and animals as cheap and disposable. Whether in the GDR or Iowa, a similar landscape emerged: vast seas of grain and oilseed monocultures; archipelagoes of concentrated, confined livestock; assemblages of exploited, disempowered workers; and watersheds of poisoned rivers and contaminated lakes.[14] Bearing this convergence in mind, the GDR's multiple environmental crises of the 1980s—the crisis of air and water pollution, the breakdowns in its farms, and the series of temporary food shortages—appear not as an aberration, but in fact as a central feature of industrial agriculture everywhere. They show that cheap food is in fact unimaginably expensive. And in the 1980s, its costs spilled over into all facets of East German life.

PIGS AND EAST GERMAN HISTORY

Communist Pigs is a history of communist agriculture looking West. It traces the rise and fall of the GDR from the perspective of one animal, *Sus scrofa*, the pig. Taxonomically, *Sus scrofa* has a genus and species all to itself. *Sus* is one of several genera—the exact number is a subject of debate—in the broader family *Suidae*, a group that includes warthogs, peccaries, and pygmy hogs.[15] All domesticated pigs descend from this elusive ancestor, the Eurasian wild boar. Thus all pigs—from the hairiest wild boar to the most scientifically-improved factory pig—are, in fact, the same species.[16] Although this book begins with one East German pig, it finds three pigs, which I call the industrial pig, the garden pig, and the wild boar. Each archetype challenges the popular memory of East Germany as a dreary, dirty, and oppressed land. Each connects the history of the GDR and Eastern Bloc to the economic and ecological transformations of Western Europe and the United States during the Cold War. Each reminds us of the active role of animals in human histories.

Animals are not static, trans-historical actors.[17] They change, and they change the world around them. While many animals lived in the GDR, including millions of cattle, chickens, and sheep, tens of thousands of house pets, and untold numbers of wild animals, no animal, save *Homo sapiens*, was more central to the arc of East German history between 1949 and 1989 than *Sus scrofa*.[18] And in this story, the industrial pig played the most prominent role. In the first two decades of postwar reconstruction—roughly 1949 to 1969—it occupied a fundamental position in the regime's uneven, halting attempts to create a modern industrial food system. Planners remade collective farms to promote grain monocultures and animal confinement. They went abroad seeking technology and expertise in vertically integrated hog production, even importing live pigs to alter the genetic makeup of their hogs. Throughout this period, East German planners measured economic progress in heads of hog and sold that message to its citizens (and Western skeptics) in kilograms of pork. By the end of the 1960s, agriculture had advanced and production levels exceeded those of the prewar period. The GDR's numbers, however, still lagged behind production in the West. Chronic problems—supply chain bottlenecks, sudden shortages, and missed quotas—persisted. Expensive machinery, agrochemicals, and massive state investment had failed to boost production to a level that could support meat exports while simultaneously feeding the people. According to the GDR's own metrics, *Sus scrofa* didn't measure up.

Then the rules of the Cold War changed. In the 1970s, four major events destabilized the postwar economic and political order, remaking global capitalism, and with it, the relationship of the GDR to the West: the end of the Bretton Woods agreement in 1971; the Russian grain deal of 1972 (which has been called the Great Grain Robbery); the Basic Treaty between the FRG and GDR in 1972; and the oil crisis of 1973. For a short period in the mid-1970s, the rate of inflation well exceeded interest rates on short-term loans—in effect making borrowed money almost free.[19] Taken together, these events produced a realignment in the conditions of global capitalism, which rebounded to the benefit of East Germany and *Sus scrofa*.[20]

Suddenly, planners had access to Western markets, and a world of "cheap things" poured into the GDR and down the gaping maws of millions of industrial pigs.[21] From the perspective of East German planners and international observers, this boosted the country's gross national product. Cheap Western grain alleviated production pressure on East Germany's collective grain and hog farmers while boosting domestic pork production. The pig population rose from nine million in 1972 to over twelve million by the early 1980s. Cheap capital, gained through loans and Soviet oil, allowed the GDR to import more Western technology, which increased production. Cheap capital underwrote an expansion of mega livestock complexes for hogs, but also for poultry and cattle.[22] Under these new conditions of world trade, the regime believed its agricultural sector would export enough pork products to gain a market share in the West. By the 1980s, East Germany's hog farms resembled those in the Iowa countryside and appeared familiar to any American agribusinessman from Hormel or Iowa beef processors. The turn toward the West saved the industrial pig in East Germany and boosted the country into the top ranks of world economies. It allowed the regime to declare it had achieved "real, existing socialism," a society where every basic need, from food to housing, work and healthcare, were met. It also planted the seeds of an ecological crisis that would destabilize the country a decade later. Until 1969, both East and West Germany shared a remarkably similar record of environmental pollution. After 1970s, their records diverged.[23] Now we know why.

If the 1970s were the decade of cheap grain and capital, the 1980s were the decade of rising costs, and the East German regime found itself trapped between the demands of its citizens and world markets. To the east, the Soviet Union announced steep cuts in deliveries of the GDR's most reliable export, Soviet oil, while to the west, creditors called in their loans as interest rates climbed and a global recession set in. The factory farms struggled to satisfy domestic demand and simultaneously increase exports. The regime faced a dilemma: should it extricate the GDR from Western markets by cutting grain imports and scale back pork exports? Or should they continue to service their debts and fulfill their contracts (following the so-called iron law of exports)? Again and again, the regime chose to prioritize the latter. *Communist Pigs* shows how the turn toward the West not only caused the expansion of East German industrial agriculture, but also accelerated the environmental crisis, driving the factory farm system to the breaking point as it struggled to satisfy Western creditors.

The turn toward the West was not just critical to the economic and political situation; it also induced transformations in *Sus scrofa*. It was in this decade that East Germany's three pigs—the industrial pig, the garden pig, and the wild boar—became most clearly differentiated and visible in everyday life. Under the new conditions of global capitalism, industrial pigs grew fatter and more numerous, but also less healthy and more destructive to the environment. In supermarkets, new economic relations boosted domestic pork supplies and also people's expectations of their government. At the same time, other pigs once thought relics of the past increased in number. Wild boars, the folkloric symbol of German nature suddenly appeared in suburban developments, collective farm fields, and state forests, becoming a public nuisance. Garden pigs returned to private plots and allotment farms, opening a porcine-shaped loophole in the regime's hostility toward these so-called remnants of bourgeois capitalism.

Communist Pigs is not the first animal history of the GDR. Annet Laue and Patrice Poutrus have written studies of animals that explore the nature of the dictatorship, society, and culture, major concerns of the first generation of historiography.[24] *Communist Pigs* builds upon these works, but expands the perspective beyond East Germany's Wall to see the country in relation to the rest of the world. Pigs show not only that the economy of the GDR bore remarkable similarities to Western economies, but also how the country became enmeshed in global transformations of markets and food production, which profoundly altered East Germany's economy and environment during the 1970s and 80s. In order to see how this happened, we need to understand how the study of animals can reframe stories about the past.

PIGS, PIG BEHAVIOR, AND ANIMAL AGENCY

Sus scrofa is a compelling animal to think with. A deserved reputation for intelligence and affinity with humans, *Sus scrofa* is the very ideal of charismatic megafauna. Orwell gravitated toward the pig for this reason, fixating on its reputation for intelligence to evoke the cunning and scheming Bolsheviks. When an editor "made the imbecile suggestion that some other animal than the pigs might be made to represent the Bolsheviks," Orwell complained in a letter to T. S. Eliot, "I could not of course make any change of that description."[25] There is more to the metaphor of pigs acting like humans. As it turns out, we all carry some pig within us. Modern genetics and molecular biology have revealed that pigs and humans share many genes and critical proteins. Our body parts are so closely analogous that pig heart valves and pig skin are used for human transplants. We share omnivorous diets and even pathogenic viruses.[26] Pig-human history is partially written in our cells, organs, and DNA.

More than any other trait, *Sus scrofa*'s supreme talent for adaptation brightly underlines East Germany's unusual role in the history of industrial agriculture. Adaptation comes in many forms. There is the variety of phenotypes (coloring, size, shape) expressed across dozens of domestic breeds. Age, sex, and climate can even effect the physical shape of a pig. This proclivity for physical adaptation has made it surprisingly difficult to determine the line between pure breeds and mixed, but also genetically between wild and domesticated pigs. Adaptation is also a feature of pig behavior, as *Sus scrofa* has established habitat on every continent except Antarctica. Carried by ships or swimming on their own, pigs have spread to remote islands and vast continents, upsetting native ecology. In a new environment, they can alter their diets and behaviors to fit the new habitat. Wild or semi-feral pigs have thrived in oak forests in Britain and on open savannah in Africa. Pigs have been kept as backyard recyclers of organic waste, as grazing animals, and as cuddly pets. Feral pigs can protect themselves from large predators—even from humans—who underestimate their speed. They regularly outcompete other wild animals for food and habitat. Second only to the Norwegian rat, *Sus scrofa* is perhaps the most successful colonizing animal in history.

Physical appearance, however, has never been a true marker of natural difference among pigs. Rather, it is an expression of *Sus scrofa*'s fluid, historical relationship to humans. The domestic hogs of early modern Europe were thinner and bristlier than their modern counterparts, as they lived mostly out of doors, fending for themselves in forest and pasture commons.[27] In the late eighteenth-century, the arrival of fatter Chinese hogs in Europe and the United States provided the genetic foundation for modern pig breeds, which depended more on a diet of grains and pulses. These "improved" pigs were meatier, had floppy ears and short legs, and were codified into dozens of

FIG I.1 "Pig Breeds." Brockhaus's Konversations Lexicon, 14th edition, 1895. F. A. Brockhaus in Leipzig, Berlin, and Wien. Courtesy of the Robarts Library, University of Toronto.

SCHWEINERASSEN.

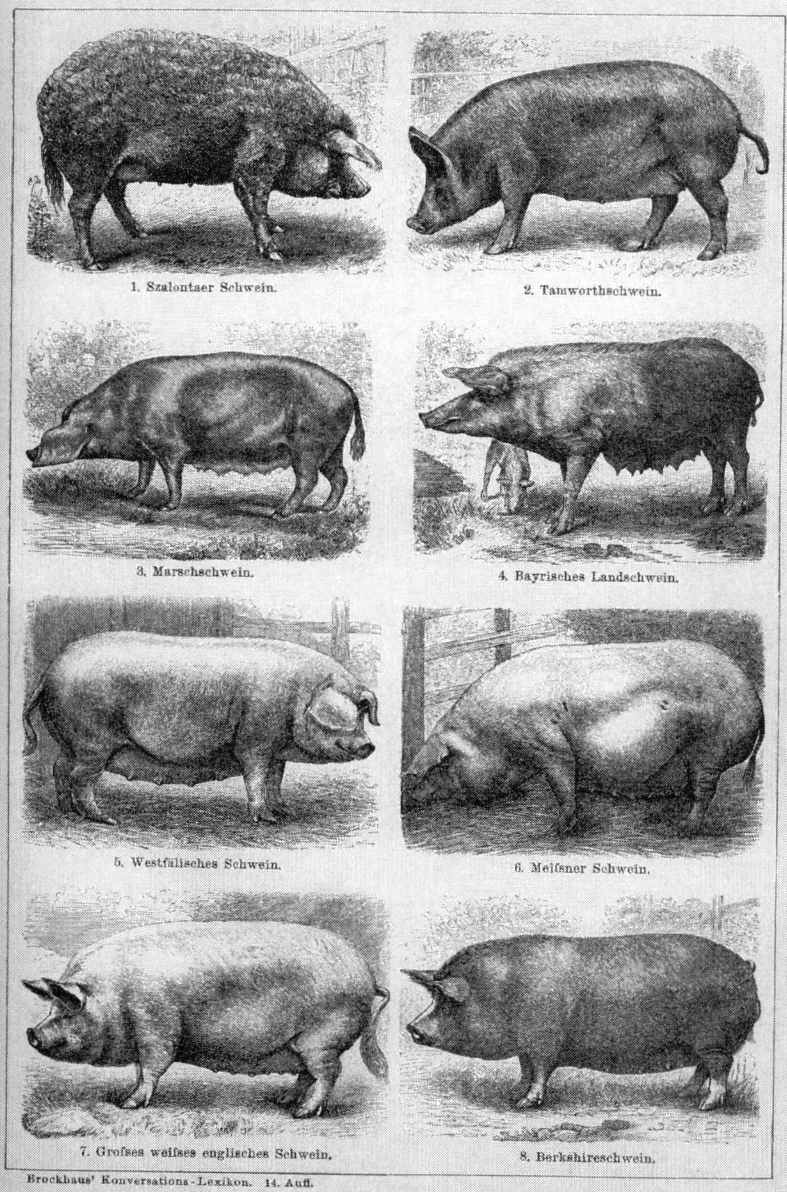

1. Szalontaer Schwein.
2. Tamworthschwein.
3. Marschschwein.
4. Bayrisches Landschwein.
5. Westfälisches Schwein.
6. Meifsner Schwein.
7. Grofses weifses englisches Schwein.
8. Berkshireschwein.

Brockhaus' Konversations-Lexikon. 14. Aufl.

distinct breeds.[28] Most amazingly, all domesticated pigs carry in their genes a store of wild boar DNA, waiting to be activated in case of emergency. If faced with hardship, piglets will access those genes and begin to revert to their wild boarish forms in only a single generation.[29] In each case, *Sus scrofa*'s physical appearance reflected the ways in which humans remade the natural world to suit their needs, and at the same time, how the pig adjusted as well.

Sus scrofa's malleability can be read in the historical record to uncover their agency, which is expressed in their bodies and behaviors. Agency is a useful concept, but also a controversial one.[30] A careless, poor conceptualization of agency risks carrying analysis toward infinite regression. If a historian is going to take seriously the agentive actions of pigs, why not birds, insects, sugar beets, or rocks? Does the acknowledgment of agency in nonhuman actors reduce the agency of actual people? If everything has agency, does that mean nothing has? The answer requires a clear definition of agency, and one that I borrow from geographer Mrill Ingram. Agency is not a zero-sum resource; it is neither experienced nor produced equally. It is an emergent and relational quality, created and observed among a given set of actors be they human, animal, or bacterium.[31] The challenge for embracing this kind of agency is to change no less than how we understand our place in the world, to see how fraught divisions between "us" and "them" have become. Perhaps it's a divide that matters less than we think. As Donna Haraway argues, rather than relying upon distinctions between things, we must pay attention to the myriad ways all things "become with others."[32]

CAPITALIST PIG, FASCIST PIG, COMMUNIST PIG?

After the Second World War, pigs and people entered into a new set of unequal relations, exemplified by the rise of the factory farm and defined by the precepts of industrial agriculture. While this new relationship emerged during the Cold War, it cannot be understood apart from the earlier history of capitalism and agriculture in the first half of the century. In the United States, *Sus scrofa* played a key role in the expansion of the country's economic and territorial empire, argues J. L. Anderson in *Capitalist Pigs*.[33] Pigs deepened the links between enslaved and free society prior to the Civil War, enabled the consolidation of capital and nature in Chicago's stockyards afterward, and served as a key commodity in the creation of the US "free trade" empire during the twentieth century. The United States was of course not the only country whose economic fortunes turned on the pig. Nazi Germany also saw the animal as vital to its own continental empire. As Tiago Saraiva shows in *Fascist Pigs*, Nazi agricultural planners developed a new standard for pig breeding, which they called *bodenständig*, or rootedness, to serve the demands of an economically self-sufficient Eastern Europe. This new and racially inspired pig aided the Nazi war effort by consuming domestic fodder sources like potatoes in lieu of costly and scarce foreign grain imports.[34] The animal also conformed to Nazi myths about pigs, Aryans, and Western civilization.

Both fascist and capitalist pigs reflected in their size, diets, and environments the ideologies of their masters, serving in one case the market, and in another *Das Volk*. In the US example, breeders of capitalist pigs selected for traits that fit the demands of national or international markets, like lard hogs in the late nineteenth century, or lean meat hogs in the twentieth. At the same time, the rise of these pig types promoted a rise in extensive markets for grain fodder, which undercut on-farm cultivation and the prosperity of whole regions of farmers on opposite sides of the globe. In this way, capitalist pigs belonged not to any country or nation, but instead to an interconnected world of capital and trade. In contrast, the Nazi's *bodenständig* pigs broke with a century of tradition in hog breeding, rejecting a global agricultural system that British and American empire had built to promote capitalist exchange, and instead selecting pigs that could flourish on the produce of sunlight-starved Northern and Eastern Europe.[35]

The GDR, although avowedly anti-fascist *and* anti-capitalist, descended from both of these histories, as did its pigs. This fact raises critical questions about the limits of ideology and ideology's relationship to animals. For example, while fascism or communism can bring a pig into existence, give it shape, and place it within a particular economic system, does that pig continue to "perform" those ideologies when the systems that created it no longer exists? Or if a pig bred for one system is deployed in another, as was the case with capitalist pigs traded to communist countries, does the pig cease its capitalist behaviors once it enters a non-capitalist society? As East Germany's pigs show, ideological distinctions are not so neat. Just as pigs reveal ideology in their bodies, behaviors, and diets, they also transform ideology—its meaning and performance—along the way. When enacted through living creatures, ideology becomes much more pliable, messy, and, as was the case for East German communism, quite brittle.

Communist Pigs shows how the East German regime pursued the industrial pig to realize the project of "real, existing socialism," only to see those same pigs advance the global conquest of industrial agriculture in the GDR, remaking the country's environment, economy, and regime in the image of the rest of the world. First originating in the US West, industrial agriculture was an ideological project based upon a capitalist configuration of nature.[36] Half a century before the founding of the GDR, "factories in the field" created unprecedented bounty for US wheat farmers.[37] By the 1930s, however, the same factories accelerated ecological chaos in the form of the Dust Bowl. The massive dirt storms that darkened American cities, destroyed farmers on the western plains, and made international headlines, were the direct result of the high cost of factory farming.[38] An analogous process would unfold in East Germany a generation later.

In their attempt to create a modern socialist society—one where all material needs, including food, would be met—the East German regime unwittingly adopted a capitalist model of development for their farms. The specific requirements of factory farming and the industrial pig—in particular, expensive technology, calorie-rich fodder, and cheap capital—drew the country deeper and deeper into world markets in the 1970s and 80s. As East German pigs feasted on rye grown in Rheinland Pfalz or wheat cultivated in Kansas, acquired with West German marks that had been initially exchanged for refined Soviet crude processed in the GDR city of Schwedt, the regime's ideological commitments to socialism weakened. Over and over, grain, capital, and oil circulated through pigs, rendering them into meat, lard, bones, and ever greater amounts of manure, which poisoned East Germany's rivers, lakes, animals, and people. In the end, the GDR had become a "state capitalist" regime.[39] As the economic and environmental costs continued to rise, the regime refused to change course, preferring to squeeze every last kilo of pork out of the system to satisfy their creditors. The environmental crisis eventually became a political one, bringing the regime close to collapse in 1982, and then permanent dissolution in 1989.

Just as the industrial pig undermined the promise of "real, existing socialism," the garden pig allowed everyday East Germans to push back, demanding access to private plots and the right to small-scale farming. Acting as a kind of star species of subsistence farming and animal breeding, the garden pig occupied the marginal spaces on the periphery of the planned economy. They included small private plots on collective farms, where garden pigs reigned, to backyard gardens and exurban garden colonies, where people grew fruits and vegetables, planted flowers, kept bees, and raised small animals like rabbits and chickens. Gardening has a long, ambiguous history in Germany, dating back to the industrial revolution, a legacy about which the government was uneasy.[40] It was only in the late 1970s, when the industrial pig had reached unprecedented levels of production, that the regime relaxed its control over these spaces and gave in to the demands of its citizens. Within a decade, however, as industrial agriculture foundered under the demands of economic planners, rampant pollution, and volatile financial markets, small-scale production became central to the survival of the entire food system. Not only did garden pigs, and thus garden farmers, prove their worth, they allowed everyday East Germans to expand the definition of "real, existing socialism" beyond a guaranteed, basic quality of life. Under socialism, everyday East Germans came to believe that they also had a right to rest, recreate, or work in a green space of one's own.

Wild boars show how some pigs were able to take advantage of an environment remade for industrial agriculture, raising questions about who and what exactly belonged in a healthy socialist natural world. More than any other pig, the wild boar—whether on medieval coats of arms or in folk tales—was the most pervasive cultural symbol of nature in Germany. For centuries, hunting had been the provenance of Germany's changing upper classes, from kings and emperors to industrialists and Nazi leaders. The regime was no different, setting aside hunting grounds, cabins, and rifles for the politically privileged. As the GDR industrialized its farms in the 1970s, the wild boar population exploded. Larger fields, simplified rotations, and a simultaneous expansion in forest cover provided the perfect habitat for wild boar populations to feed, breed, and expand their range. East Germans who hiked into the woods to commune with nature found themselves facing not only marauding wild boars but mountains of toxic industrial pig manure trucked from overflowing sewage lagoons on massive collective pig farms. There was no place in the country untouched by pigs.

East Germany's three pigs ask us to reconsider how "real, existing socialism" actually worked and why it ultimately failed. All three animals flourished under this system yet also acted as agents of ecological chaos. The global exchanges that made the industrial hog so successful were also responsible for the acceleration in environmental degradation and the resulting economic chaos. *Communist Pigs* offers snapshots of an environment and ideology undergoing complicated and disruptive change. When looked at from this perspective, the multiple, malleable bodies of *Sus scrofa* reveal a history of state socialism in flesh and bone.

In *Animal Farm*, the seven revolutionary commandments are painted on the side of the barn so all the animals can learn them. The seventh commandment—the most important—is "All animals are equal." Eventually, the ruling Stalinist pigs amend the seventh commandment to read, "All animals are equal but some animals are more equal than others." With apologies to Orwell, East Germany's pigs are more than equal to the task of connecting the history of the GDR to the environmental history of the world in the twentieth century. They reveal the quixotic project of cheap food and its high costs. It's a story much greater than the size and even lifespan of the GDR. It is nested in different layers of history: of Germany, of communism, of the Cold War, but also of the United States, of capitalism, and of animals—especially pigs.

The industrial pig, the garden pig, and the wild boar lead us through the rise and fall of East German industrial agriculture but also foreshadow changes around the world where largescale schemes for industrial agriculture are imposed without regard for people, animals, or environments. In East Germany's collapse, they point to an underappreciated source of the regime's political instability. Our guides will be three pigs, each a sterling example of one of the most adaptable animals on earth. Indeed, should the hot, polluted world of capitalist nature come to an abrupt end, the surviving *Sus scrofa* might prove to be the more equal animal. In Orwell's book and this one, pigs make their own history.

CHAPTER ONE

WHEN PIGS COULD FLY

NINETY MEN AND WOMEN LINED THE EDGE OF THE TARMAC AT Schönefeld Airport in March of 1970, awaiting a charter flight from Ljubljana, Yugoslavia. The flight was entering its fourth hour as it neared East Berlin, and its passengers were agitated, stressed, and hungry. None had ever traveled in an airplane before. But response teams on the ground were prepared, ready to provide food, water, medical care, and transportation. Even in an era when air travel had become the norm across Europe, this flight was unique. None of its passengers was, in fact, human. The East German response team was waiting to receive thousands of specially bred sows and boars. They were the final contingent on the last leg of an astonishing, unprecedented airlift of hogs to the German Democratic Republic.[1] In East Germany, pigs could fly.

Once on the tarmac, the plane was swarmed by workers from East Germany's preeminent hog producing complex at Eberswalde. Urgency was necessary, as the workers knew that travel produced deadly levels of stress in hogs, activating the so-called halothane gene, a recessive trait that was a marker of improved hybrid breeding but also caused some pigs to drop dead when under duress.[2] The workers quickly divided the animals into eight tractor trailers for a hurried drive north to the city of Eberswalde. It was there that East Germans hoped to produce a new type of pig, one that would carry desired traits for meat and fat content yet withstand the arduous conditions of the factory farm. If successful, this new industrial pig would allow the GDR to produce pork of a quality and quantity to rival any country in Europe. The last planeload of breeding pigs to land was the culmination of a tremendous undertaking. Between November 1969 and March of 1970, the East Germans imported over nine thousand breeding hogs from Yugoslavia to jumpstart their own industrial hog sector. The pig airlift involved eighty-eight direct flights, all overseen by top management at the Yugoslav firm Emona in Ljubljana and its East German counterpart in Eberswalde.[3]

The flight symbolized a new phase in the economic development of East Germany, and the end of an era. For the last twenty years, the government had pursued a top-down industrial revolution in the countryside, punctuating agricultural reforms with a series of coercive, intermittent, and often violent collectivization drives. Their ultimate goal was food self-sufficiency, or as the slogan put it, the transformation of the country from "an import land to an export land." By the mid-1960s, the government had succeeded in creating thousands of collective farms (*Landwirtschaftliche Produktionsgenossenschaften*, or LPGs), but not an export dynamo. In response, they turned their gaze abroad, eventually settling on the model of factory farm in 1964. This new model prioritized the development of state owned farms over collective farms and became the basis for complexes like Eberswalde. It is the story of Eberswalde—of the origins, installation, and spread of the factory farm—that serves as one half of the story of this chapter. The other half belongs to the animal that lived inside, the pig.

Was Eberswalde built for the pig or the pig for Eberswalde? Which came first, the factory farm or the factory pig? During the Cold War, experts in swine science might have argued that their technology was tailored to fit the pig. In their eyes and textbooks, swine scientists claimed *Sus scrofa* was preternaturally disposed to factory life. They pointed to an array of traits suited to production quotas and accounting balance sheets. Every sow could bear up to two litters a year, each producing between eight and fourteen piglets. Each one of these five-pound piglets could grow into a massive 270-pound feeder hog in the space of six months. The final product was an animal that was mostly meat, yielding between 72 and 76 percent of its total body mass (cattle and sheep yield between 62 and 63 percent and 50 and 57 percent, respectively).[4] East German scientists shared this assumption, and at Eberswalde, they would eventually raise up to 190,000 hogs over a five-month period. This meant that nearly four hundred thousand hogs would live and die in the factory complex over the course of a single year. For East German agricultural planners in search of high-yielding animals, the pig was second to none.

A closer look at the origins of the factory pig, however, reveals that this animal had to be created, or even better, retrofitted to suit the factory farm. As its managers discovered, Eberswalde proved to be a sensitive instrument. The factory farm demanded strict control of people and pigs alike, coordinating hog life cycles, agricultural labor, and heavy machinery. It needed new types of inputs, like vast amounts of calorie-rich grain in ready supply. It also required a new type of pig. In the 1960s, the classically bred East German pigs—products of the age of "improvement"—struggled under factory life. Looking to international standards, and with the aid of the Yugoslavs, East German agricultural scientists specified new traits for the country's latest generation of factory hogs. Spending their entire lives within the factory farm environment, these new pigs had to best their predecessors in weight gain, litter size, and lean-meat percentage, while developing higher tolerance for disease and confinement. Breeders called this new constellation of characteristics "vigor." In form and function, they represented a new kind of hog entirely, one I call the industrial pig.

The industrial pig was ideology made flesh. More so than other animals, pigs' bodies are expressions of human culture. Their physical form, growth rate, and meat type are relatively sensitive to environmental conditions and human manipulation. Where pigs live, what pigs eat, and how many pigs live together all produce changes in the animal. During the Cold War, farmers, breeders, and politicians united behind the ideology of the factory farm, which left observable traces in the very bodies of pigs. Yet it was not an ideology that belonged solely to East Germany or the socialist states of Eastern Europe. The new vigorous hybrids that populated the massive sheds at Eberswalde followed a global trend. They inherited a genetic stock that was drawn from centuries of breeding in Europe, the United States, and Asia that was then reformulated again at mid-century for the factory farm by capitalist and socialist animal breeders. In short, they were bred for the factory farm, a powerful new ideological organization of animal agriculture, which demanded a transformation of all living things—animals, plants, and people—that encountered it.[5] During the Cold War, East German planners, Iowan agribusinessmen, and US and Soviet trade representatives all fell hard for the factory farm and the pigs that lived and died within it.

In the decades after World War II, East German industrial pigs performed as historical artifacts of a revolution in industrial agriculture. This drama unfolded slowly across Europe and the United States between 1945 and 1975. The 1970 charter flight symbolized the convergence of two systems and one ideology in a single animal. East and West, capitalist or communist, modern agriculturalists agreed on a common vision of industrial meat production. And changes in *Sus scrofa*—to its body and its environment—mirrored changes between the East German regime and its farms, Western Europe, and the United States. East Germany's flying pigs beat a path for the factory farm in Eastern Europe. From the United States and West Germany, to Yugoslavia, the GDR, and the Soviet Union, the industrial pig routinely crossed Cold War borders.

EBERSWALDE AND INDUSTRIAL AGRICULTURE IN THE TWENTIETH CENTURY

Eberswalde sits sixty kilometers northeast of Berlin near the Polish border. The place was fitting for such a facility: *Eber* means boar and *Wald* means forest. Since the middle ages, the city's crest featured two razor-backed, sharp-tusked wild boars beneath an acorn-laden oak tree, a tribute to the time when semi-feral hogs fed themselves on the mast fruit, or nuts, of hardwood trees.[6] In 1967, East Germans planners broke ground at Eberswalde. In what would become the largest such complex in the world, one of four in the GDR, Eberswalde's leaders pioneered a troika of industrial facilities that disaggregated the entire life cycle of the hog into three branches of inputs and outputs: grain, breeding, and meat.

The site was vast. At a port along the Oder-Havel Canal, construction crews erected the first factory—an array of office-building-sized grain silos, concrete warehouses, and processing equipment, collectively known as a *Kraftfuttermischwerk* (concentrate fodder-mixing works, or KFME). The factory received and prepared grain fodder from across the GDR and Europe, tailoring the nutrient and mineral percentages in each feed sack to the specific age and sex of every stage of hog life in the factory.[7] The KFME shipped thousands of tons of fodder across the canal to the next facility, the swine breeding and fattening combine (SZMK) Eberswalde, the destination of the airlifted pigs. The facility's technicians, specialized in every stage of the hog's life cycle, cared for the breeding, birthing, weaning, and fattening of pigs. Here, Eberswalde's directors showcased the latest technological advances in industrial hog keeping, such as automated feeding, gestation crates, and

FIG 1.1 Exterior photograph of Schweinezucht und Mastkombinat (SZMK) Eberswalde. Courtesy of Horst Schneider, Private Collection/wirtschaftsgeschichte-eberswalde.de/agrarwirtschaft.

concrete slated flooring. Once at full capacity, half a million hogs would pass through the SZMK's doors every year.

As fast as Eberswalde could assemble hog life, it could disassemble it even faster. After separating the breeding stock from the meat hogs, managers shipped market-weight pigs north to the edge of town to the *Schlacht und Verarbeitungs-Komplex* (SVK Eberswalde/Britz), a turnkey-ready slaughterhouse and packing facility built in the mid-1970s entirely by a West Berlin firm, Consult, with Western technology, capital, and expertise.[8] SVK Eberswalde was designed not only to supply the communist side of Berlin with pork, but also to produce an assortment of meat products for export to West Germany and the Soviet Union. Supporting the complex was a vast infrastructure of rail lines, canal ships, cranes, irrigation ditches, sewage lines, and waste treatment plants, all tying together the disparate phases of porcine birth, life, and death. Eberswalde was truly a city remade for hogs.

FIG 1.2 Exterior photograph of Schlacht- und Verarbeitungskombinat (SVK) Eberswalde/Britz. Courtesy of Horst Schneider, Private Collection/ wirtschaftsgeschichte-eberswalde.de/agrarwirtschaft.

The complex at Eberswalde symbolized a new East Germany in the 1970s: technologically advanced, outward facing, highly educated, and modern. Eberswalde modeled agriculture and food production for the rest of the country, employing expertise, demonstrating managerial control of labor, promoting specialization, and utilizing the latest technologies and machinery. It coordinated every stage of pork production, processing, and distribution—the very model of vertical integration in agriculture. In addition to pigs by the millions, Eberswalde drew East Germans by the thousands, from scientists and researchers across the East German university system to workers and laborers from every branch of farming and industry. Eberswalde's directors built schools, training centers, grocery stores, concert halls, parade grounds, and single-family homes and provided vacation rentals for its workers. They invited agricultural students and scientists from around the world of socialist "friendship" to study the latest methods and techniques.[9] Most astounding of all, the complex pulled the GDR into the top ranks of world pork producers. According to the US Department of Agriculture (USDA), between 1970 and 1982, East Germany became the largest pork producer in the entire Eastern Bloc outside of the Soviet Union, doubling its output by 1981. Perhaps even more impressive, the GDR surpassed renowned Western European pork producers, such as Denmark and the United Kingdom, over the same period, while falling just short of the much larger Federal Republic of Germany and France.[10] To the people who built it, Eberswalde was an East German success story.

Eberswalde's origins directly contradict the popular image of communist agriculture as backward and inefficient. For much of the Cold War, and in the histories written since, Western critics decried this facility, and communist agriculture more generally, as symptomatic of the heavy hand of the totalitarian state. Eberswalde, and other complexes like it, unleashed epidemics of pollution and disease upon East German people and animals, bolstering further criticisms of communist agriculture. How then do we reconcile Eberswalde's gleaming modernity with the crises it caused? Where does Eberswalde fit within the global history of industrial agriculture at mid-century? And how do we account for the vast disconnect between the perception and reality of the Hog City?

The answer partially lies in the Cold War, which in the words of Mary Nolan "inflected everything from the arms race and economic development . . . to economic and cultural interactions."[11] This was just as true for agriculture and economic development. As the protagonist in the 1982 novella *The Wall Jumper* said about divided Berlin, "It will take us longer to tear down the Wall in our heads than any wrecking company will need for the Wall we can see."[12] The wall in our heads still shapes our understanding of the communist states of Eastern Europe. Industrial agriculture not only crossed this wall, but it also predated the wall's construction. Attention to the origins of industrial agriculture in the first half of the twentieth century reveals how the model came to the GDR in the first place.

Industrial agriculture was a US export, originating at the turn of the twentieth century in the US West. There, imperial conquest, capitalist expansion, and dumb luck transformed once "worthless" land into sites of capitalist speculation.[13] Upon this space, wealthy landowners, USDA agents, and regional bankers conspired to remake American agriculture around a new industrial model, first developed for commercial wheat farms. Boosters urged farmers to adopt heavy machinery, immense economies of scale, and a Taylorist division of labor, which they underwrote with a widening network of transportation infrastructure and inexpensive inputs, such as cheap land, low-cost labor, and low-interest loans. The new farmer was no longer a worn-down laborer, but instead a factory boss, managing the farm through account ledgers. While farmers (owners) devoted increasing amounts of their time to tracking sales, credit, and workers' wages, a cadre of permanent managers supervised seasonal laborers in the farm's day-to-day operations.[14]

The industrial ideal gained traction and further articulation during the First World War, when European demand surged, pulling more and more landowners into wheat production and the disciplining sphere of credit and debt. After the war, European demand collapsed, sending overleveraged American wheat farmers to ruin yet leaving the new model intact. Within a couple of years, agricultural production surged again, this time in Texas, Oklahoma, New Mexico, and Colorado and in the western states of Montana and California.[15] Paradoxically, the more farmers the industrial model ruined, the more entrenched the factory farm became. Even in the midst of the Great Depression, the industrial model endured.[16]

The rise of industrial agriculture in America did not go unnoticed in Europe. In the 1920s, German and Soviet economists became particularly obsessed with the growing power of United States agriculture and industry. Seeking to find the keys to American modernity, they studied the astounding productive power of US factories, assembly lines, and managerial practices. Germans analyzed terms like *rationalization*, *Fordism*, and *efficiency* and topics like the emblematic River Rouge plant and the American factory worker.[17] Debates over the "American model" also extended to agriculture and, in particular, the low cost of food. In his 1925 book *Economic America*, the German industrialist Carl Köttgen focused on principles of American industrial production such as standardization, mechanization, and mass production. But he also examined agriculture. For Köttgen, America's natural advantages, particularly its soil, were a major factor in producing real incomes that were 70 percent greater for American workers than German counterparts—a fact that he believed was made possible by the low cost of food in the United States.[18] Unwilling to take the capitalists at their word, a fourteen-man delegation of German trade unionists traveled to the United States that same year to inspect labor conditions in factories, oil fields, and farms in the Midwest and Northeast.[19] Importantly, the labor union delegation visited small family farms in areas least likely to have adopted industrial practices by 1925. If the trade unionists had gone to the West or Southwest, they would have seen another face of American agriculture labor—the share cropper, tenant farmer, and migrant laborer—and heard firsthand from management the new gospel of the factory farm.[20]

This is not to say that there were no German antecedents to the industrial model. The market and commercial revolutions of the nineteenth century left their mark on the region's landscapes, animals, and states. Works like Daniel Albrecht Thaer's *Principles of Rational Agriculture* and Johann Heinrich von Thünen's *Isolated City*, and Justus von Liebig's discovery of the principles of soil chemistry, were produced in the context of an emergent German commercial agriculture and centralizing states at mid-century. Innovations in scientific husbandry convinced the state of Saxony to create the world's first agricultural experiment station in 1850. The US government would write this model into the 1862 Morrill Act, which established the land grant university system. In eastern Prussia, the commercialization of the sugar beet industry upended class structures, creating not only a new class of large estate landowners—the infamous Junker class—but also racialized groups of rural laborers.[21] At the turn of the twentieth century, German agriculture certainly had many of the ingredients of the industrial model but had yet to put to work the full constellation of features—like mechanized equipment, a Taylorist division of labor, and economies of scale—in the service of a new, capital-intensive mode of mass (food) production. All that changed in the 1920s. The sudden and rapid adoption of the tractor that decade was a leading indicator of the arrival of the American industrial ideal in Germany.[22]

Fascination with American modernity was not confined to Germany's Social Democrats, industrialists, and trade unionists. It was also observed from the fringe. Adolf Hitler saw the United States along with the Soviet Union as the two greatest threats to German dominance. Hitler looked to the United States' continental empire as a model, and he particularly admired the genocide of Native Americans that had helped create it. Hitler believed the United States' tremendous economic advantages were a result of what the German dictator called the "size of America's internal market" and its "wealth in purchasing power but also in raw materials."[23] The Americans had not just inherited this vast internal market, but in fact had taken it through blood and conquest in order to create "living space." At the outset of his campaign for an eastern European empire in 1941, Hitler declared that the Volga River would become Germany's Mississippi, an internal transportation network that would connect the German metropole to its colonial frontier. And just like the American West, the German "East" would be made possible through the removal of the "racially inferior" natives in favor of Aryan settlers. "Here in the East a similar process will repeat itself for a second time as in the conquest of America," Hitler declared. "Europe, not America, will be the land of unlimited possibilities."[24]

Hitler believed that industrial agriculture would play a key role in the resettlement of the East, a model that was distilled in the industrial killing complex of Auschwitz. Heinrich Himmler envisioned the slave camp, when originally constructed, as serving as the premier agricultural research station for an eastern empire, holding ten thousand inmates for manual labor.[25] As it evolved, Auschwitz was made to serve IG Farben, the Nazi economy's most important corporate collaborator in its quixotic synthetic rubber campaign. IG Farben used thousands of Auschwitz inmates to work in their greenhouses growing *kok-sagyz*, a dandelion-like plant that Farben scientists believed could be the raw material for plentiful synthetic rubber.[26] At Auschwitz, a modern white supremacist dream was made reality, with the factory farm at its center.

At the other end of the political spectrum, the Bolsheviks also looked to US agriculture, believing the Americans had discovered the future with the factory farm. The Soviets vowed to emulate it. In the late 1920s and early 1930s, the Soviet Union established exchanges with American agronomists and farmers in order to share their agricultural expertise. Experts from California's land grant universities joined businessmen and factory managers from America's largest corporations in delegations to the Soviet Union. The American experts helped plan large-scale development projects in the Soviet Union, like a four-hundred-thousand-acre wheat complex and a specialized hog-breeding unit. In return, the Soviets sent their own delegations to the United States. The exchanges stopped when Stalin tried to accelerate agriculture modernization and simultaneously break peasant resistance through forced collectivization in 1931. Americans (along with all foreigners) were expelled. The American agricultural experts roundly condemned Bolshevism and resented their expulsion, but many begrudgingly continued to admire the size of Soviet wheat fields.[27]

Whether expressed in the Nazis' belief in American *Lebensraum*, the German industrialist obsession with managerial expertise, or the Soviet infatuation with monocultures and economies of scale, all agreed that American industrial agriculture was the way of the future. Yet missing in the European analysis was the role of nature in America's productivity, largely because all of them—including the Americans—took nature for granted. As Jason Moore and Raj Patel argue, capitalist production has depended on the separation of all things into the realms of nature and culture since its emergence in the fifteenth century.[28] In this process, what Europeans codified as natural, capitalists deemed free and therefore available for the (violent) taking. As capitalism expanded, it drew in these "free gifts" from the periphery to the center of economic production. In agriculture, think of the "empty" land stolen from Native Americans, the labor of "racially inferior" slaves, the "traditional" childrearing and caretaking of women, or the fossil fuels and organic fertilizers "harvested" from the subterranean world or captured from the air. This idea was so pervasive that it transcended Cold War divisions. As an East German collective farm chairman named Ulli Wruck explained to his fellow directors in 1966, "We cannot expect Nature to simply bestow her gifts upon us. . . . We must wrest them away ourselves!"[29]

These "free gifts," however, were never free, as capitalist ecology tended toward inevitable exhaustion and degradation. Part of the power of this system was its ability to hide the damage and destruction it caused through short-term fixes. Capitalists introduced new machinery, bought larger farms, hired migrant laborers, and purchased synthetic fertilizers with cheap credit to cover up declining crop yields, market busts, or soil exhaustion. The illusion could only last so long. The more these fixes pushed off a true reckoning, the more severe the degradation and the more lasting and widespread the damage. The United States had its first reckoning with the high cost of this system with the Dust Bowl. The massive dirt storms that darkened American cities and ruined farmers across the western plains were the direct result of the high cost of factory farming. As Donald Worster argues, "A common thread [ran] . . . from the tractors, banks, and large-scale commercial farming to black blizzards, abandoned farms, and Okies."[30] Capital drove them on toward expansion, exhaustion, and collapse. At the heart of both the Dust Bowl and the European fascination with US American agricultural productivity was the very same capitalist configuration of nature—one that would only grow more influential in the postwar era. Far from avoiding these pathologies, the GDR would follow in the path of its predecessors.

POSTWAR OCCUPATION AND THE RECONSTRUCTION OF AGRICULTURE

The GDR was heir to this history of industrial agriculture when founded in 1949. Although avowedly anti-capitalist, its early leaders, from First Secretary Walter Ulbricht to President Wilhelm Pieck, fetishized the power of industrialization and thus the capitalist configuration of nature it demanded. In the 1950s, Ulbricht in particular made "the building of socialism"—a euphemism for crash industrialization—the country's central economic priority, exhorting East Germans to "rebuild," or "*bau auf!*" the ruined country, factory by factory.[31] Over that decade, the regime constructed new industrial towns, like Eisenhüttenstadt (a massive steel-mill combine), Schwedt (an oil refinery for Soviet crude imports), and Halle-Neustadt (a chemical refinery, dubbed the City of Chemical Workers), solely for the expansion of the economy's heavy industry base.[32] Authorities targeted agriculture for a similar ideological redevelopment in the same decade. Land reform would pave the way.

The future of the region that would become the GDR was uncertain in 1948. For one, it was unclear if "Soviet" Germany would become socialist at all. At Yalta in 1945, none of the Big Three—Stalin, Roosevelt, or Churchill—imagined the shared occupation of Germany that they had just agreed to would become a permanent division.[33] As late as December 1948, Stalin cautioned the German communists that the region was not ready for Sovietization, saying, "No people's democracy yet . . . not direct interventions, but zigzag . . . cautious policy."[34] The immediate postwar confronted all of the Allies with a series of existential crises. Socialism would have to wait.

The Soviet military, along with German communist leaders, struggled with several crises between 1945 and 1949. Hitler's war had created endemic food shortages in Germany, where calorie consumption fell to 860 per day as of June 1945, well below the wartime high of 2,445 calories per day of 1941. It had destroyed half of the region's housing stock, including 75 percent of Berlin's apartments, and left an estimated twenty million Germans homeless. In retribution for Germany's crimes, the governments of Eastern Europe expelled every German in their respective countries. Nineteen million Germans flooded back to occupied Germany from lands recently conquered by the Nazis, including Czechoslovakia, Poland, Hungary, and the Soviet Union. There were millions of non-German displaced peoples, including POWs, slave laborers, and concentration camp survivors moving around Europe. Soviet soldiers raped hundreds of thousands of German women, which led to the births of 150,000 to 200,000 "Russian babies" in 1945–46.[35] It was in the context of these humanitarian crises that the Soviets carried out a major round of land reform in 1946.[36] The program, which redistributed ownership rights and property among new owners, expropriated 3.3 million hectares of Junker (aristocratic) property in the Soviet Zone, giving land to landless farmers, agricultural workers, and ninety thousand German refugees recently arrived from Eastern Europe.[37] In total, the program created 855,624 farms out of the ruins of the Nazi state.[38]

The 1946 reform served a dual purpose, part ideological and part practical. In the former case, the Soviet program upended the Nazi-era structures of rural society, namely the immense *Reichsnährstand* (RNS) and immense commercial estates of the Junker landlords. The RNS was a Nazi institution that once controlled all agricultural production within Germany. It determined prices and production levels in every village, town, and regional city, coordinating food production between farmers and processors. It was the "largest single economic unit in the world" at its height, controlling 40 percent of the German workforce, 25 percent of GDP, and winning thirty billion reichsmarks in sales every year.[39] The RNS and large Junker estates were synonymous with Nazi power, and so the Soviets targeted them for ideological purification, as expressed in the infamous 1946 communist slogan, *Junkerland im Bauernhand!* or "Junker lands in the farmers' hands!"[40]

There were practical reasons for land reform as well. It offered a low-cost way to establish subsistence-level production—though little more—among an occupied and defeated population. Many German farmers were shocked to discover that their new farms came with ownership rights and little else. They had neither the tools nor the machinery, let alone barns, livestock, or a market to sell to. The plots were uneconomical.[41] Many were quickly abandoned as their owners sought work in industrial centers or in the West. Other farmers made the best of a challenging situation, reconstituting the cooperative farming organizations of the prewar years, such as the Farmers' Association of Mutual Assistance, in order to pool labor and machinery.[42] The 1946 land reform was highly disruptive. It provided a backstop against famine and rooted thousands of new German farmers in the countryside. At the same time, it carried out indiscriminate ideological retribution, displacing thousands of other Germans, many of whom were Nazis. Most important for the future East Germany, it was the first step toward industrialization, preparing the ground for eventual development.

The Soviet Union was hardly the only country that saw land reform as a prerequisite to modernized agriculture. The United States promoted land reform assiduously throughout the first decades of the Cold War. Many American civil servants believed that land reform was a "powerful lever of social change" available to "both sides."[43] In Germany, US Treasury secretary Henry Morgenthau proposed a dramatic land reform program that divided the country into three "agricultural and pastoral" states, while removing all industrial infrastructure and equipment. It was a plan so drastic that British diplomats called it a new "Carthaginian Peace," akin to sowing the German economic soil with salt.[44] While the US land-reform scheme was never enacted in Germany, the United States pushed harder for land reform across postwar Asia. In Korea, Major General William Dean redistributed Japanese farmland to more than five hundred thousand farmers in 1948, while also reducing Korean rents on expropriated property.[45] In 1950, the United States attempted to prop up the Chinese nationalists by sending money to help with the redistribution of property across the Yangtze Valley.[46] The United States also adopted a technological route to land reform, which took the form of dam construction.[47] Like communist planned economies, dam projects required a strong central authority, which expropriated and redistributed property across the project area, ordered the construction of farms and factories, built roads, housing, and schools, and dictated production levels.[48] Modeling their programs on the Tennessee Valley Authority, American officials sold TVAs around the world to promote state power and capitalist development.

For ideological reasons, the United States was much less willing to pursue land reform at home. In the United States, the conventional wisdom of the early 1950s asserted that communist land reform, and collectivization in particular, was anathema to an American system of free enterprise and family farms. In Congress, lawmakers emphatically claimed that collective property was antithetical to a prosperous agricultural sector. As the 1960 Family Farm Income Act declared, the US government supported "the family system of agriculture against all forms of collectivization."[49] Such ideas had deep roots in the American Cold War psyche. In 1955, Walt Rostow, Kennedy's modernization theory architect, noted in his own history of communist agriculture that the Soviets had managed to produce a system of "unproductive and enslaved agriculture" where collectivization had led to only "sagging" agricultural output.[50] Even dyed-in-the-wool left wingers like the labor and economic historian Lewis Corey dismissed collective agriculture as "quack magic."[51] As Cold War competition deepened, collective agriculture became the foil against which many Americans defined their farms and economies.

This perception, while deeply held, obscured a more important truth: US policymakers saw land reform as an essential prerequisite to the industrial development of agriculture. Between 1950 and 1970, the US government encouraged agricultural modernization through indirect coercion of American small holders, allowing credit and debt to drive tens of thousands to abandon their farms. This was especially the case in Iowa, whose Cold War farming history finds a striking parallel in the GDR. Both regions embraced the same underlying logic of industrial agriculture—that the state should promote large-scale operations to produce enough cheap food to feed the expanding working classes and to trade around the world. Both regions simplified the ecology of their rural spaces to fit a new agroecology of grain, oilseed, and livestock.[52] Both regions also experienced a remarkably similar set of symptoms: displaced people, consolidated farms, and expanded fields, all with a hidden plague of environmental degradation. In East Germany, they called this program "collective agriculture"; in Iowa, they called it "agribusiness." From this perspective, we can argue that Iowa and East Germany share the same history.[53] It was in these regions that capitalist and communist agriculture took on a new, shared character.

COLLECTIVIZED AGRICULTURE IN THE GDR

In the Soviet Zone of Occupation, the poor situation in agriculture meant that land reform could not stop with the expropriation of Junker land in 1946.[54] The East German regime knew agricultural development would have to continue. In 1952, the Socialist Unity Part (SED), the ruling party of East Germany, ordered the next phase of land reform—collectivization—to begin. The plan, however, was not a one-size-fits-all model. Instead, it promoted three distinct types of LPGs, with varying degrees of collectivization. Type I combined agricultural land, while individual members were allowed to keep their animals and machinery. In Type II, members pooled their machinery and land but kept their animals; in Type III, farmers transferred all the means of production to the cooperative.[55] In 1952, the regime promoted the program relentlessly, relying on a mixture of political agitation, government incentives, and direct pressure. The "agitators"—low-level party functionaries—tried to convince farmers of the economic and political advantages of collectivization, while offering state aid to the villages that joined. They gave farmers who joined LPGs preferential access to heavy machinery, two-year tax exemptions, free advice from extension stations, supplies of low-cost seeds, consumer goods, and cheap credit.[56] Collectivized farms also introduced new management roles to the farm and created a bureaucracy to negotiate with planners over quotas, bid for raw materials, inputs, and machinery, and schedule planting seasons. Management also organized labor brigades, set wages, instituted a forty-hour week, and offered vacations to all employees.[57]

Collectivization's land reform unfolded alongside Ulbricht's increasingly draconian austerity program for the country at large. In 1952 and 1953, the regime ordered price increases and dramatic cuts to state subsidies for families, while raising factory worker quotas across the board. Ulbricht believed industrialization had to come before all else in the GDR. In response, thousands of people simply left the country. Sixty-eight thousand East Germans fled in the first half of 1952, followed by another two hundred thousand over the following calendar year.[58] Only a nationwide uprising on June 17, 1953, and its immediate suppression by Soviet tanks, eased austerity. In response, Ulbricht retracted price and production increases and repealed the forced collectivization clause from the 1952 law, allowing collective farms to voluntarily dissolve. Many did. Across the GDR, 564 LPG's, with over twenty-three thousand members, disbanded in 1953.[59] In the subsequent months and years, Ulbricht moved away from direct political pressure, favoring the soft power of political and economic incentives to get farmers to consolidate their holdings—all the while retaining full collectivization along industrial lines as the ultimate goal.[60]

Ulbricht's collectivization interlude lasted until 1959. In 1960, he wanted to finish the job. Despite Soviet premier Nikita Khrushchev's objections, on January 27, Ulbricht declared a renewed goal of "full collectivization" as the primary task for the country. During this Socialist Spring, the regime sent *Volkspolizei*, Stasi agents, and other security officials to buttress collectivization efforts and quell possible resistance.[61] Cadres of agitators poured into the countryside, pressuring independent farmers to join. The size of the groups and the use of loudspeakers and floodlights intimidated many.[62] Yet political pressure was just a part of it. As one farmer remembered:

> All sorts of people traveled around to the villages to try to generate enthusiasm and win over every last farmer for the "cooperative path" as they called it. They did this in all sorts of ways. First through friendly conversations. Of course, there must have been cases . . . where there were people who say, "If you don't cooperate you will land in Siberia," and things like that. Such thing may have happened, but there was never any legal basis for it. . . . People did try to blackmail farmers a bit and to drive them into a corner. They also used enticements such as promising a car during a time when cars were so scarce.[63]

Not all farmers were willing to comply, and many turned their rage, despair, and anger at the East German state upon their animals. Farmers stabbed and maimed their own livestock rather than see healthy animals fall under the control of an LPG. Others resorted to subtler yet equally cruel methods. The *Volkspolizei* described farmers who poisoned their own animals' fodder with shards of glass, nails, chemical fertilizers, and pesticides. The most common act of resistance, however, was the intentional slaughtering of livestock, especially calves and piglets. In 1959 and 1960, neglect, injury, and premature slaughtering of privately owned livestock resulted in the death of 1.25 million pigs, 200,000 cattle, and 115,000 sheep.[64]

The Socialist Spring came to an end in 1961, and the regime declared victory. The number of LPG IIIs had increased to more than twenty-thousand, covering 85 percent of the country's arable land and counting over a million members. By this measure, it was a massive success. It was also a costly one. In the first half of 1960 alone, the GDR lost 14,700 farmers; they joined more than two hundred thousand other East Germans moving west in the same year.[65] On a much smaller scale, yet not less horrific, some farmers committed suicide, despairing over the loss of prized animals and property. Only the construction of the Berlin Wall in August of 1961 brought the *Republikflucht* to halt.[66]

Despite their resistance and the considerable hardship inflicted upon many East Germans, collectivization cannot be understood as solely an ideological struggle over property rights. Some feared they would become like the Soviet peasants who, in the words of one farmer, "own nothing anymore and have to go around dressed in rags."[67] However, many objected to collective agriculture for practical reasons related to farm labor and income distribution: poorly organized work shifts, low production levels, diminished profits, and unequal compensation. All contributed to a widespread sense of dissatisfaction with the regime.

The greatest source of frustration for collectivized farmers was the persistent shortage of agricultural laborers. While flight to the West played a role, the majority of East Germans emigres were older educated professionals, not typical farm workers. Traditional farm labor was back breaking, low skilled, and poorly paid. It was work suited to the young or socially marginal. East Germany's rural labor shortage, however, was a product of the success of its reconstruction. Like many industrial countries after the war, East Germany created a surplus of more highly paid jobs in factories and urban centers. In addition, recruitment to the *Volkspolizei* and National People's Army pulled

East Germany's rural youth out of the countryside en masse during the 1950s. They fled not for their lives, but in search of better pay and a higher standard of living.[68] While coercion certainly played a role, ideological resistance does not explain the challenges facing collective agriculture over the succeeding decades. These are understood more easily when placed within the broader structural transformations in the societies of postwar Europe. The changes occurred in the East *and* the West. New farming technology such as combines, automated feeders, and spray irrigation offered farmers everywhere the chance to permanently bypass labor shortages. These trends in East German agriculture of 1960 had a clear counterpart—in Iowa of the 1950s.

IOWA AND AMERICAN AGRIBUSINESS

American agriculture emerged in the postwar period in a stronger position than that of East Germany's, but it too was transmogrified in the wake of the Great Depression. Hunger and economic devastation brought the regulatory arm of a strong centralized state into the countryside in the 1930s. The Agricultural Adjustment Act of 1933, one of the first laws of the New Deal, moved economic planning to the center of national agricultural policy. Through price supports and production controls, the American government sought to end the volatility in commodity markets and to protect incomes for farmers and food prices for urban consumers through what became known as "planned scarcity."[69] Not all were convinced that this was the best method. In particular, a new cadre of pro-business administrators, led by Ezra Taft Benson, came to power in the 1950s, taking over the USDA and setting new priorities for agriculture. Under Benson's leadership, the USDA shifted its focus from protecting farmer incomes to promoting American agriculture as a single industrial sector of the economy. The best way to promote rural well-being, they argued, was to make farming more efficient. Driving marginal small holders from the land was one part of this program. Iowa was not immune to this process. In place of "planned scarcity," the USDA promoted a new term, *agribusiness*, in American agriculture.[70]

First coined in 1955 by Harvard Business School professor and former assistant secretary of agriculture John H. Davis, agribusiness stood for a merger of industrial capitalism with agriculture.[71] It grew out of a nineteenth century view of agricultural progress, embodied in the land grant agricultural university, where science and mechanization would force modernization in the name of the national economy.[72] In the context of the Cold War, boosters defined agribusiness in direct opposition both to communist agriculture and the New Deal. Supporters, like Eisenhower's secretary of agriculture Ezra Taft Benson, lambasted the same New Deal institutions that had saved farms across the country as a fifth column in the American heartland. American agriculture, he believed, would benefit from the removal of price supports—which violated the free market—and instead wanted the state to raise incomes through the expansion of markets for agricultural commodities. This change in policy, Benson argued, would require greater economies of scale and more technological fixes, and it would thus transform the food chain, from the farmers' fields to the processing facility, and on to the supermarket and the dinner table.[73] Between New Deal "socialism" and the laissez faire capitalism, boosters believed agribusiness offered a third way for the family farm.[74]

Speaking for the USDA, Davis argued that agribusiness was not a political choice but rather the inevitable outcome of technological progress. In a 1957 speech, he argued, "It is important that we understand that vertically integrated agribusinesses are the result of forces generated by technology. . . . Resistance to change that is inevitable merely serves to build up problems and add to human suffering."[75] In his vision, however, it would be private corporations, rather than the federal government, that would provide price stability, because of economies of scale. The large processors of agribusiness would offer a steady supply and large distributors a stable price structure for farm produce. Critics attacked Davis as protecting monopolists or, even worse, mimicking the gigantism of Stalinist agriculture. When pushed to clarify how the American farmer would resist becoming a "mere cog in the machine as now exists in Russia," Davis balked, saying only that the American farmer was a "better type" of man, less vulnerable to such forces.[76] Despite the anticommunist rhetoric, Davis's neologism, agribusiness, unwittingly described a system with ironic similarities to communist agriculture.

During the 1950s and 1960s, agribusiness rolled through rural Iowa like a market-based collectivization drive. While it could not send the army with bayonets to force small producers to sell, consolidate, or expand, the ideology of agribusiness wielded debt and credit to compel farmers into increasingly tenuous positions relative to their neighbors and the market. If they chose to go their own way and sell directly to the market, they would have to take out large annual loans to pay for the heavy machinery, fertilizers, hybrid seeds, and irrigation systems that promised to boost yields. If their gamble on the season busted, debt was rolled over into the next year or it rolled over the farmer. While some won this gamble, most did not. There was another option. Rather than subject themselves to price volatility, many Iowan farmers chose to offload their financial risk and contract directly with major food producers. In return for a guaranteed payout, these farmers renounced control over their operations, as they were impelled to produce a narrow list of commodities at specific prices. As a result, contract farmers lost freedom over their daily operations and found themselves vertically integrated into an industry controlled by enormous corporations, such as Campbell Soup, Ralston Purina, and Perdue.[77] Either option compelled an expansion in field size and winnowing of owners and crop diversity and often ended by forcing people off the land.

Unsurprisingly, the number of Iowa farms dropped precipitously in the postwar period. While the total cultivated land in Iowa remained around twenty million acres, farms expanded, increasing from 160 acres in 1950 to 239 by 1969, as the number of farms dropped from 205,000 in 1945 to around 135,000 in 1970.[78] Nationally, a similar story unfolded, as the average farm grew by a third while farms fell in number by a similar percentage in the first fifteen postwar years.[79] Agribusiness also left noticeable marks in Iowa's fields and grocery stores, creating a boom for commodity producers and a bust for diversified farmers. Whereas in 1920, Iowans produced an assortment of grains, crops, and livestock—"horses, cattle, chickens, corn, hogs, apples, hay, oats, potatoes, cherries, wheat, plums, grapes, ducks, geese, strawberries, pears, mules, sheep"—by the 1950s, three commodities dominated—soy, corn, and pork.[80] As the latter two commodities soared—growing from 462 million bushels to 858 million bushels, and 10.9 million hogs to 17.8 million—the state's vegetable, fruit, and secondary produce production collapsed.[81]

Iowa's rural exodus rivaled East Germany's, with a net migration out of the state of nearly six hundred thousand people between 1940 and 1970.[82] This mirrored national transformations as well. While there were 7.2 million farmers in the United States in 1931, the number had fallen all the way to 2.8 million by 1973.[83] And like East Germany, a labor shortage among young workers contributed to the collapse in labor-intensive vegetable and fruit production. By the 1970s, Iowa teenagers were staying in school longer than in the 1950s and 60s, and when they graduated, they moved to cities for higher paying jobs. Producers offset the labor shortfall with expensive machinery—a fix that relied on commodity production more amenable to automation and economies of scale. As one Iowa farmer reasoned, "Good dependable hired help is getting harder to find. So, the best way for me to sidestep this labor bottleneck was to mechanize my feeding [for his hog farm]."[84] Machinery and automation, driven by the ideology of agribusiness, concentrated agricultural production onto fewer larger farms. Of the roughly three million American farms in operation in 1973, the vast majority were four hundred acres or fewer in size. Two percent of all farms were responsible for more than a third of all agricultural production in the country.[85] The trend was noticeable everywhere. A 1960 aerial survey of Indiana's farms described the same process of "enlargement of fields by the removal of fences and the joining of small fields."[86] The most productive Iowa farms were also the largest and the most mechanized.

The triumph of agribusiness transformed rural America. By the 1970s, agribusiness had pushed so many people off the land that rural communities became heavily dependent on food produced elsewhere. In echoes of East Germany's Socialist Spring, small towns withered away. Financial ruin broke up families. Lost livelihoods pushed farmers into depression and in some cases to suicide.[87] The similarities with communist land reforms were not lost on the great critic of agribusiness, the writer and farmer Wendell Berry. "I remember, during the fifties, the outrage with which our political leaders spoke of the forced removal of the populations of villages in communist countries. I also remember that at the same time, in Washington the word on farming was 'Get big or get out'. . . . The only difference is that of method: the force used by the communist was military; with us it has been economic . . . the attitudes are equally cruel, and I believe that the result will prove equally damaging."[88]

Despite the similarities, agribusiness escaped much of the condemnation directed at collective agriculture, largely due to the centrality of technology to the ideology. Quoting from a 1970 *National Geographic* article on American agriculture, Berry highlighted the author's wonderment at the high-tech paradise, where "an incredible parade of machines are at work today on U.S. farms: acre-eaters . . . self-propelled combines that permit a man to ride in an air-conditioned cab to harvest a crop of corn that used to take a crew of 80 hands."[89] Technological utopianism, however, was not unique to the United States. The article could have just as easily come from the mouth of Leon Trotsky, who once described the tractor as a "cultural tugboat," pulling the Russian peasantry into modernity. Like their capitalist counterparts, communist leaders saw technology as a "value-neutral" force for democracy and progress during their time.[90] This belief pervaded both communist and capitalist attitudes about agriculture. Consider that in the same year Davis was preaching the gospel of agribusiness—1957—the Communist Bloc celebrated the successful launch of the Sputnik satellite. Thrilling to this real, demonstrable victory of socialism over the West, the East German politician Fritz Selbmann linked the triumph of Sputnik to agriculture. "He who is first in sending an earth satellite into space," Selbmann declared, "will also . . . outstrip capitalism in the production of meat and fat."[91] If technology could lead the communists into space, then surely it could make them first in the supermarket.

In the 1960s, East German leaders looked abroad for new, large-scale technologies like the animal factory farm. Parallel systems moved closer together during that decade, as the GDR purchased machinery and animals on an international market. Importing this new technology was one thing. Making it work was another. As Eberswalde's directors and East German planners eventually discovered, the factory farm demanded the reorganization of not just labor, but also the land surrounding it and the animals within it.

FROM YUGOSLAVIA TO EAST GERMANY

The factory farm was not indigenous to the GDR; planners shipped it in, piece by piece. Following the construction of the Berlin Wall in 1961, Ulbricht's devotion to autarky (or economic self-sufficiency) weakened. In its place, the first secretary pushed economic reforms intended to diminish central planning in the economy.[92] In agriculture, this meant turning away from collective farms as the central site of industrial development. Instead, Ulbricht looked abroad to growing transnational networks of agricultural expertise and trade that encompassed the Socialist Bloc, non-aligned countries, and Western Europe. Planners kept intact the principles of "collective agriculture"—like managerial control, heavy machinery, economies of scale, specialization—but designated state-owned farms and research institutions responsible for their development and dissemination. These new factory farms served as models to the rest of the country and brought the GDR into closer orbit with the West, and indirectly, the United States.

The GDR's agricultural development in the 1960s depended on cooperation with new partners, chief among them Yugoslavia. East Germany's prior patron, the Soviet Union, had grown less interested in the domestic affairs of the GDR since the construction of the Berlin Wall.[93] The following year, a series of international and domestic crises—the Cuban Missile standoff and collapse of the Virgin Lands Campaign in particular—threw Khrushchev's regime into chaos, turned Soviet politics inward, and caused decreased economic support for the GDR.[94] The East German regime suddenly found itself free to follow its own path. Czechoslovakia and Hungary offered opportunities for collaboration, but it was the non-aligned, independent Socialist Federal Republic of Yugoslavia that pushed East Germany's factory farms into the future, through indirect access to the technologies of American agribusiness.[95]

Yugoslavia occupied a unique position in the geopolitical organization of Europe. Ever since the Stalin-Tito split in 1948, the United States had attempted to entice the Yugoslav leader, Josip Broz Tito, through money, arms, and trade, to withdraw from the Warsaw Pact and open the country to Western tourists, businessmen, and perhaps military influence—the culmination of the so-called wedge strategy.[96] The Americans saw the outreach to Yugoslavia as a way to undermine the Soviets but also as an opportunity to bring the Yugoslav economic system closer to the capitalist world.

For the Yugoslavs, foreign investment offered resources to rebuild after World War II and civil war. From 1953–65, the country experienced rapid economic growth, largely fueled by borrowing from the United States, the United Kingdom, and Italy.[97] In this period, the Yugoslavs established relations with state-run banks, like the American Export-Import Bank, to finance the import of millions of dollars' worth of industrial machinery, food, and raw materials. The bank's credit also made new trade relations possible, many of which were centered on agricultural production. For example, in 1960, the Yugoslav National Bank and trade ministry Jugoexport financed a sale of Yugoslav beef, fed on imported grain from the United States, to the American military in West Germany and Italy.[98] American agribusiness saw potential in the socialist country. In Zagreb, a million Yugoslavs visited an exhibit on the American supermarket in 1957—anticipating the famous Moscow Kitchen Debates between Nixon and Khrushchev two years later.[99] Americans believed the supermarket technology would revolutionize agriculture from the checkout counter to the farm. The Yugoslavs were less certain about the potential of self-service grocery shopping and refrigerated display cases. They did, however, continue to purchase modern agricultural equipment and technology from Italy, West Germany, and the United States.[100] American agribusiness was happy to oblige. In 1965, John Deere sold four thousand tractors to the Yugoslavs.[101]

Occupying a central position between capitalist and communist blocs, Yugoslavia acted as a link to the West for Eastern Europe.[102] Beginning in

1964, the Yugoslavs entered talks with the GDR to expand economic cooperation across a range of industries. In the realm of agriculture, the GDR's State Purchasing Committee (Staatlichen Komittee für Erfassung und Aufkauf) and the Yugoslavian agricultural firm Emona took the lead, organizing two exchanges—one in Berlin in November of 1964, the other in Ljubljana in January of 1965—for representatives to tour the host country's agricultural facilities.[103] At the time, Emona was at the forefront of world agricultural practices, and the East Germans were in awe. In the last week of September in 1966, Ulbricht traveled to Ljubljana to see the *Agrokombinat* for himself. The agro-combine managed three integrated complexes—one for chickens and eggs, one for beef, and one for pork. The complex spread out from the northern outskirts of Ljubljana across three thousand hectares of land, employing more than two thousand people. Ulbricht marveled at how it took nearly twenty-minutes to drive from one part to another.[104] Emona, whose name came from the ancient Roman settlement that predated the Slovenian capital, was particularly proud of its poultry operation, which was the first stop on Ulbricht's visit. The model facility included "twelve flat, aluminum sided halls," each containing fifty thousand chickens. Visitors were required to observe the strictest protocols over hygiene, changing their clothes and wearing white coats upon entry.[105]

The East German guests marveled at the advanced automation that controlled everything from feeding to heating and lighting. The Yugoslavs even had machines that sorted, counted, and stamped eggs.[106] After the poultry facility, Ulbricht toured the modernized barns where fifty thousand pigs were produced in just six months. Ulbricht saw the vast works of Emona as evidence of the realization of socialist modernity, and he wanted it for the GDR. "When we have a strong, modern socialist economy, we will be met with the proper respect of the capitalist economies," declared Ulbricht during a rally in the Ljubljana city center.[107] Shortly thereafter, Emona sent construction crews and technical advisors to the East German city of Königs-Wusterhausen to begin construction of the country's first factory farm for Goldbroiler chickens.[108]

The 1965 agreement over Königs-Wusterhausen was the first phase of a plan to build factory farms all over East Germany. Emona also oversaw the construction of three industrial cattle facilities in the regions of Rostock, Magdeburg, and Brandenburg; five broiler operations in the same first two regions plus Potsdam, Halle, and Karl-Marx-Stadt; and three more combined egg and broiler facilities, one of which would be in Königs-Wusterhausen.[109]

In a 1967 interview with *Neues Deutschland*, the party newspaper of the GDR, Emona's director of technological development, Franz Dolinar, said he had driven so often between Berlin and the sites of the new mega-facilities that he knew the roads in the region of Frankfurt am Oder "like the back of my hand," or in German, like his *Westentasche* (coat pocket).[110] It was at this time that Dolinar had begun planning the construction of one of two industrial hog breeding and fattening facilities in the small town of Eberswalde.[111]

Workers broke ground in the fall of 1967. Construction unfolded in two phases. The Yugoslav engineers designed a pavilion-style site for the swine breeding and fattening combine, a collection of more than forty specialized long barns arrayed in four rows and connected by covered walkways and roads. Inside each barn, steel and concrete-spalted flooring lay underneath hundreds of animal stalls, covering two to four waste canals that led out of the building. Above the stalls, a network of metal walkways allowed workers to move easily within the barn, while a large automated feeder (resembling a Zamboni ice-rink machine) crawled along elevated tracks, pouring a nutritionally calibrated fodder ration into troughs below.[112] Emona also constructed

FIG 1.3 Interior of fattening barn, SZMK Eberswalde. Courtesy of Horst Schneider, Private Collection/wirtschaftsgeschichte-eberswalde.de/agrarwirtschaft.

a drainage system and a waste lagoon; a control room for thermostats, ventilation, and fodder distribution; heating and cooling systems for the piglet barns, breeding rooms, and fattening sheds; twenty-one grains silos with capacity for seven days of fodder; a power station; a cadaver house; a factory restaurant; a training school; and other ancillary structures. The GDR agreed to pay Emona the equivalent of $19 million for the complex, which they paid out in three distinct currencies—East German marks, gold reserves, and dollars.[113] On October 7, 1969, the regime declared SZMK Eberswalde open for business.[114]

WHITHER THE INDUSTRIAL PIG?

In October 1969, Eberswalde had everything it needed except the right animal. This too Emona provided. There in the contract was a special provision for the delivery of "necessary animal materials" from Ljubljana, or 4,200 breeding sows and 114 breeding boars, all between six and ten months old.[115] A later revision to the terms would double the number of pigs to more than nine thousand. In November of 1969, the first phase of the pig airlift took off from Ljubljana, bound for East Berlin.[116] At the same time, the directors of Eberswalde signed a contract with East Germany's animal breeding organization (VVB Tierzucht), which was made up of more than sixty-four different facilities and labs, to provide an addition 3,500 breeding pigs in the nearby town of Münchenberg. Together, Emona's and Münchenberg's pigs would form the foundational breeding stock for SZMK Eberswalde, which would swell in population to more than sixty-five thousand pigs by the end of 1970.[117] The total operation was costly, requiring extensive planning and cooperation. It was also a big risk, as the chartered flights very easily could have killed their live cargo. Why did the Eberswalde's directors need Emona's pigs? Why couldn't the East Germans just use their existing breeding stock? By the 1960s, animal technicians and breeders knew that not all animals could withstand the factory farm, let alone flourish in them.

Around the world, the transition to industrial confinement was anything but seamless. It put enormous stress on pigs by restructuring their lives. First, factory farms reorganized the biological and seasonal rhythms of porcine life, as European and North American farmers turned increasingly to lifecycle feeding and housing to manage their hogs.[118] Under this system, farmers separated their pigs by age, weight, and sex into specialized housing units. Sows, for example, spent their gestation days in a separate pen to minimize fighting between sows and to protect against spontaneous abortion. Shortly before giving birth, sows were moved to a farrowing shed, where they would remain with their piglets for the first few days after birth. Generally, piglets stayed with their mothers until about six weeks, when they were moved into a new facility with weaner pigs to transition from milk to dry feed. Until they reached market weight, pigs were kept according to their age groups, moving pens and changing feed types with each successive stage. In general, the age classification correlates strongly with weight, going from weaner pig (maximum weight twenty to forty pounds), feeder pig (forty to seventy pounds), shoat pig (not sexually mature, weighing 150 to 260 pounds), to hog (weighing more than 220 pounds).[119] The real innovation of lifecycle housing was that it allowed farmers to breed their sows twice a year, making year-round pork production possible. In this way, the factory hog farm eliminated seasonality in pork production and marketing.[120]

The industrial ideal may have satisfied the demands of the planned economy, but it also created horrific living conditions for the pigs themselves, who suffered. Writing in 1978, the Romanian agronomist A. S. Terent'eva noted that only eight to ten of the classical breeds were "suitable" for industrial production.[121] Breeders now sought animals that could withstand this environment, and so they turned to heterosis, or hybridity. Unlike the "improved purebreds" of the nineteenth century, the new pigs were selected for traits that could tolerate the engineered environment and flourish under the conditions of factory farm life. For meat hogs, breeders wanted rapid weight gain, higher tolerance for disease and confinement, and greater percentages of lean meat in their final weight. Breeders prized hybrid sows for other qualities; they recovered quickly from labor, they conceived at higher rates, and they carried more litters over their lifetimes. For breeders, one word came to denote the suite of specialized skills and traits in the new factory hog—*vigor.*[122]

East German hog breeders began experimenting with hybrid hogs in 1965, working with the traditional German meat types—white pigs like the long-bodied German Landrace and *Edelschwein* (improved pig).[123] Hybridization began when breeders crossed two traditional breeds to form the "mother" line of the new hybrids and then stabilized the desired traits of this second generation—like a high percentage of lean meat—by breeding with another pig, usually a boar from a specialized pig breed, such as a Mangalitsa, Duroc, Hampshire, or, most typically, a Piétrain. This final cross transferred more traits, including the boar's feeding capacity, fertility, and resistance to stress, to the third generation.[124] The process required years of replication, reproduction, and patience. The East Germans were hardly alone, for wherever factory farms existed, a hybrid breeding program did also. By 1978, more than 90 percent of pigs in the United States were hybrid animals. In the United Kingdom, the number was 60 percent, in Holland it was 70 percent, and in Sweden it was 25 to 30 percent. In 1970, the Soviets designated hybrid breeding the center of their own economic transformation, experimenting with more than 120 crosses, drawing on both global meat types and their unkept regional varieties.[125] At the same time, the world's hog population exploded, doubling in the twenty years between 1954 and 1974, from approximately 350 million head to around 700 million.[126] At the height of the Cold War, the vigorous hybrid pig marked a new stage in its history, rapidly growing in number yet declining in genetic diversity.

As with other advancements, the United States was the first to breed hybrid hogs, beginning in the 1950s, a shift driven by US consumers' health concerns. At the time, an array of studies decried the dangers of consuming animal fat. In place of lard, public health authorities recommended vegetable cooking oils and artificial fats. The pork industry rushed to adjust. During the 1950s, meat packers such as Hormel collaborated with land grant colleges, such as Ohio State University and Iowa State University, to develop lean pork.[127] The shift from lard to lean meat inaugurated a top-down transformation directed by public/private collaboration between industry and public universities. The Americans were hardly alone in this preference. In a 1972 swine science textbook, East German experts expressed the socialist desire for healthier cuts of pork. "Today lean meat is in demand. New quality specifications have been derived to increase the meat percentage of pigs to a maximum, while keeping fatty tissue to a minimum."[128] At the Tenth Farmers' Congress (Deutschen Bauernkongress) in 1968, East German farmers passed a resolution declaring lean pork the primary meat type of all pork production within two years.[129] While consumer preferences ostensibly drove the research into leaner hogs in the United States and East Germany, the new animals also reflected the requirements for industrial facilities that produced them. Efficient fodder digestion, faster weight gain, larger litters—these too were the markers of the new industrial pig.

In the GDR, the Leicoma pig was the first hybrid to achieve breed status, and it emerged directly from the Eberswalde system.[130] Under the supervision of Gunther Nitzsche, scientists at the state's swine breeding program (VVB Schweinezucht) started to develop this new line in 1971.[131] Nitzsche

FIG 1.4 Leicoma boar with caretaker. Courtesy of Deutsches Schweinemuseum e.V.

created a first-generation cross of a Landrace and a German Sattleschwein, a local breed that was black with a white belt around its chest. Up until that point, the first generation of hybrids had failed to exhibit productive traits that exceeded those of existing commercial breeds. Part of the problem may have been the uncertain provenance of the genetic stock. Many Eastern European hybrids possessed indefinite lineage, the likely legacy of a lax adoption of standardized record keeping in the Soviet Union since the 1930s.[132] In the case of the Leicoma, Nitzsche claimed that the Soviets forced his staff to cross the second generation with a boar line of Estonian Bacon hog from the USSR, which likely possessed uncertain heritage. With tongue in cheek, Nitzsche referred to this mistaken cross with the propaganda slogan, "To learn from the Soviets is to learn victory." Unable to trust the line of the Estonian Bacons, Nitzsche rebelled. He persuaded officials at Eberswalde to exclude the Estonian Bacon in further trials and instead to replace its genetic stock with the better-known American-bred Duroc. Between 1971 and 1986, breeders worked to stabilize the hybrid line as a separate and distinct breed. It was referred to throughout experiment results and planning reports as Line 250. Since it was a result of collaboration between facilities in three major districts of Leipzig, Cottbus, and Magdeburg, the pig received a new name—LeiCoMa—and was adopted in most of East Germany's pork facilities.[133] Yet it was just as much a product of Cold War exchanges with Yugoslavia, West Germany, and the United States—a truly transnational animal.

It took years of breeding and selection to create an animal as successful as Leicoma, but in 1976, East German planners claimed victory for the vigorous pig and the factory farm at Eberswalde. In 1971, only 16 percent of the country's breeding sows were hybrids, but by 1976, the number had risen to nearly 84 percent. Pork production numbers were even more impressive, growing from 993,000 tons of pork to 1.1 million tons over the same period. Planners cited rising fertility, increased weight gain, and higher quality meat—the desired traits of hybrid programs everywhere—as a direct result of this program, with the secretary of agriculture highlighting the success of Line 250 for praise in the same report.[134]

By the mid-1970s, the GDR's industrial revolution in agriculture was triumphant, embodied in the Leicoma pig and the immense complex at Eberswalde. The report glossed over other problems, like irregular grain fodder supplies, bottlenecks in production, and factory hygiene. Yet Eberswalde's managers knew what type of pig would yield best to the industrial system erected over the course of the 1960s and early 70s. As the breeding director of Emona put it in 1967, "We already know what the best kind of meat hog is: it's the 'three-cornered pig.' Small head, little fat, good cutlets and big juicy hams."[135] Or as the breeders at Hormel, University of Iowa, and Ohio State, might have understood them, "lean hogs."

CONCLUSION

Between 1950 and 1970, East Germany witnessed a complete reconstruction of its agricultural system. From the ruins of the Nazi regime to the disruption of the Socialist Spring, a generation of GDR farmers were pummeled by political force and technological change. By the 1970s, however, social upheaval had largely subsided, and a new era of industrial production was ascendant. East Germany's farms, backed by strengthening economic and political ties with its Eastern Bloc neighbors, took on the emergent industrial practices and forms prevalent throughout the United States.

Walter Ulbricht also believed the countryside was poised to break out from the economic malaise of the past two decades and perhaps gain some economic independence from the Soviet Union. His efforts culminated with nine thousand industrial pigs trotting across the tarmac in East Berlin. Bred for the factory farm and world agricultural trade, the hybrid hog was the penultimate step in the GDR's transformation into a world economic power. There was one last barrier: the autarkic trade relations of the Eastern Bloc. So Ulbricht sought another route to the West. On June 25, 1970, the first secretary informed the deputy chairman of the Soviet Council of Ministers, Nikolai Tikohnov, of his new plan. "We get in as much debt as possible with the capitalists, up to the limits of possible," he told Tikohnov.[136] Ulbricht's plan was to use Western capital to develop the industrial base and to eventually transform the GDR into an exporting powerhouse. Unfortunately for Ulbricht, his East German rivals got wind of this apostasy and made a move against him. Erich Honecker, with the backing of Leonid Brezhnev, used the rise in East Germany's debt—2.2 billion marks as of 1970—to oust Ulbricht in December of that year.[137]

The fall of Ulbricht brought neither the issue of borrowing nor the problems of industrial agriculture to an end. On the contrary, both would continue to expand. The industrial pig and the industrial system built for it made demands that the GDR could not meet on its own. Export to the West became the name of the game. Over the next decade, Honecker would borrow sums of money Ulbricht never dreamed of in pursuit of cheap Iowan corn, Soviet grain, and Western technology. At the same time, the regime sent more and more products of East German agriculture, like Eberswalde hams, back into the West.

Honecker's gamble was clear. In the words of one of Emona's managers, "The scientific and technical performance of today depends on the economic efficacy of tomorrow!"[138] During the 1970s, the economics of tomorrow came crashing down on the factories of the day. In the meantime, engineers would need to continue working. As Meta Recnik, the chief technological engineer of Emona's egg hatchery, told *Neues Deutchland* in 1967, "What you see here will be changed shortly. We'll make it bigger. And still easier."[139] The collateral for all this debt, technological modernization, and the means to pay it off would rest heavily on the shoulders of the new hybrid industrial pig.

CHAPTER TWO

THE GREAT GRAIN ROBBERY AND THE RISE OF A GLOBAL ANIMAL FARM

ON DECEMBER 20, 1972, THE LOUISIANA-BASED BULK CARRIER *OGDEN Willamette* docked alongside an enormous concrete pier in the Black Sea port city of Odessa. On board were thirty-six thousand tons of American and Canadian wheat. Within minutes, a gang of forty Soviet stevedores swarmed aboard, dragging long hoses connected to American-made Vac-U-Vators—forklift-sized vacuum machines—to suck every kernel out of the cavernous holds and into waiting boxcars. It took four days to unload the *Ogden Willamette*. Two more American grain carriers, the *National Defender* and *Overseas Joyce*, arrived in Odessa over the next ten days, and the unloading continued round the clock. All told, the three carriers delivered nearly 150,000 tons of American wheat to the Soviet Union.[1]

The ships were the vanguard of an American merchant marine fleet crossing the Atlantic to deliver twenty million tons of grain and oilseed produced in the United States and Canada. The *Ogden Willamette*'s cargo was in fulfillment of an unprecedented trade deal completed in July 1972 between the Cold War superpowers—a deal so massive that at the time it was the largest commercial transaction by volume in world history.[2] Millions of tons of corn, soy beans, and canola seed were included, but wheat made up the bulk of the deal, an amount equal to approximately 80 percent of the United States' annual wheat consumption and more than 30 percent of its annual production.[3]

The deal was so large that it allowed the Soviets to corner the global grain market, in effect driving up prices everywhere. For a time, it made the communists running the Soviet Union the world's leading capitalist speculators. Within months, American consumers in supermarkets and politicians in Congress were hit with sticker shock. By July 1973, corn and soy prices had doubled while wheat had more than tripled. Rising grain costs cascaded into livestock and meat production too, as cattle and hog prices surged.[4] Overall food prices rose 15 percent in the United States.[5] At a Pathmark supermarket in Greenvale, New York, a crowd of angry housewives demanded an explanation for the price spike from the store's manager. "It's the Russians," he protested to the mob in the parking lot. "They bought all that grain, you know."[6] The problems, however, extended far beyond the checkout line. Six months after the initial deal, three quarters of the twenty million tons still sat in grain elevators and silos across the United States, while freight companies, grain traders, and longshoremen fought over scarce railcars and barges.[7]

In the midst of the Cold War, American politicians wondered how the Soviets had so effectively tricked them. "We were snookered," Richard Nixon declared in the summer of 1973, ignoring the fact that his administration had organized the deal in the first place.[8] Journalists recast the twenty-million-ton grain deal as a heist, the Great Grain Robbery. Calling it a robbery obscured the fact that the Soviets had cut the deal fair and square. As George Shanklin, an expert at the USDA, remarked, "The Russians were very clever. They were able to buy large quantities without bidding the price up. I give them credit for being very good capitalists."[9]

How did the Soviets get the drop on their superpower rivals? Part of the answer lay in the United States' low estimation of Soviet agriculture. Over the previous fifty years, beginning with Stalin's forced collectivization in Ukraine, Soviet agriculture had acquired a reputation as chaotic and famine prone. But 1972 was not 1933. The Soviets had made significant advances in grain production, harvesting three times as much wheat as they had a decade earlier, all in the service of industrial livestock production. When Soviet trade representatives arrived in New York City in the summer of 1972, they came not to negotiate for bread, but for meat.

The Soviet entry into the global grain market signaled a new phase in an ongoing global revolution in industrial agriculture. As we have seen, from Iowa to East Germany, industrial boosters transformed livestock farming on the inside of every barn. Just as significantly, they remade agriculture on the outside as well. Where once farmers raised crops and animals adapted to local landscapes, markets, and climates, they now pushed a new kind of global commodity farming, an agroecological transformation that Tony Weis calls the "meatification" of global diets.[10] This process promoted the cultivation of a narrow set of feed grains (wheat, rice, maize) and oilseed (canola and soy) monocultures alongside confined livestock facilities.[11]

At the center of this transformation was the vigorous industrial pig. As an individual, the animal had been bred to tolerate the harsh conditions of factory life. Yet when amassed and confined by the thousands, the industrial pig exerted a new kind of force upon the world that could disrupt markets, reorganize landscapes a continent away, and, eventually, drown cities in effluent waste—all to satiate the demands of their digestive tracts. The industrial pig flourished on a specific calorie and protein-laden diet, dietary demands that required increasingly large quantities of land devoted solely to feed production in a volume that few countries, including the GDR, could meet on their own. In pursuit of this end, governments all over the world squeezed out local and regional producers of livestock and grain to make way for vertically integrated giants such as Cargill, Continental, or Eberswalde that could supply volume to meet demand on a global scale. In order to see how the GDR fit into this transformation, this chapter leaves the industrial pigs of East Germany for the Cold War world of grain, trade, and capitalism.

While East Germany's agricultural transformation became inextricably linked to "meatifcation," it was not inevitable. It was made possible by a series of economic and political shocks—which included the Russian grain deal—that upended Western capitalism and world order in the 1970s. In 1971, the Nixon administration undid the Bretton Woods system, the 1944 set of economic agreements that created the institutions and rules of Western capitalism, and established the dollar as the global reserve currency.[12] In the near term, the system's end introduced volatility and upheaval to Western economies. The year after the grain deal, 1973, was even more disruptive, as the Yom Kippur War between Israel, Egypt, and Syria ended with victory for Israel, an embargo of oil sales to Western Europe and the United States, and a rising tide of inflation throughout the world. In this tryptic of economic shocks—Bretton Woods, the Russian grain deal, and the oil embargo—the three major "cheaps" that underwrote Western capitalism—capital, grain, and oil—underwent dramatic, sudden revaluations.[13] With the rules of Western capitalism uncertain, new players could participate. So entered the Soviet Union, Latin America, the Eastern Bloc, and of course, the GDR.

The 1970s was not just the era of economic shock. It was also the era of détente. While multilateral diplomatic agreements on nuclear weapons, borders, and human rights snagged the headlines, agriculture cleared the way. Détente opened the door to the Eastern Bloc and thus new economic agreements with the West, many of which concerned agriculture. Following the dictates of meatification, communist and capitalist leaders were eager to sign trade deals and short-term loans that swapped the most important inputs of industrial agriculture. In a matter of years, new flows of credit, capital, grain, oil, and meat stitched together the world's agricultural land into a new industrial logic of a global animal farm.

The drive toward global commodity production was uneven, especially in Eastern Europe. On the one hand, the efforts to build an integrated industrial agriculture did lead to a brief flourishing of East German, Soviet, and other Eastern Bloc economies. At the same time, financial innovations weakened the control that East Bloc regimes had over their economies and people. By the end of the decade, cheap grain and cheap credit were distorting the balance sheets of Eastern Bloc economic planners, driving state socialism to the brink of insolvency. Across the East and West, economic events moved swiftly, leaving many like the angry Pathmark shoppers struggling to make sense of how the price of meat in Long Island was connected to the standard of living for Soviet stevedores in Odessa.

A PIG'S-EYE VIEW OF DÉTENTE

The Great Grain Robbery turned the Cold War world topsy-turvy. The normally well-oiled machinery of capitalist trade now groaned under the logistical weight of moving twenty million tons of grain. From Houston and New Orleans to Pascagoula and Mobile, railcars sat for days waiting to be unloaded. Instead of a ten-day round trip, railcars on the Illinois Central Gulf Railroad crawled from grain elevator to port and back again in fifteen to twenty days. Senator Carl Curtis called it "the most serious breakdown in transportation in the history of the United States."[14] Yet here were American shippers struggling to fulfill their delivery quotas, like communist enterprise managers in the midst of a Five Year Plan. The Soviets, meanwhile, appeared not only as master manipulators of markets, but as a maturing economic powerhouse. The deal demanded that American politicians dramatically reevaluate the state of communist agriculture.

Initially, contemporaries missed the significance of the grain deal due to other world events. The year 1972 was pivotal in the Cold War.[15] In February, Nixon flew to China to normalize relations. In December, West Germany and East Germany signed the Basic Treaty, codifying the postwar division of Germany into two states, but one nation. Yet the biggest headline of the year came that May from the American and Soviet summit in Moscow, when the superpowers signed the Agreement on Basic Principles of Relations, followed in short order by the first Strategic Arms Limitation Treaty (SALT I) and the Anti-Ballistic Missile Treaty.[16] The 1972 and 1973 agreements opened a period of normalization in the Cold War.

Détente was not a single event but a long chain of agreements and negotiations, carried out between the world's three superpowers, as well as by the small powers of Europe. In newspapers and on television, the age of East-West Détente seemed to play out in airport arrivals with senior leaders waving from the tops of airplane stairs or in stuffy press conferences in the curtained ballrooms of heavily guarded palaces. The talks were mostly about submarines and MIRVs or trust and verification, but a surprising number of the détente agreements and a great deal of the money involved agriculture.[17] Since the 1950s, the first Soviet-American exchanges involved agriculture, as was the case with Khrushchev's obsession with Iowa corn or the less well-known story of the Soviet-bred bull, the 1956 International Grand Steer Champion PS Troubadour.[18] While controversial issues like missiles and third-world conflicts could abruptly end a summit, the Soviets and Americans could always talk about agriculture.

The Russian grain deal emerged from this history. It started as a small confidence-building measure by the Nixon administration. Within months, however, agricultural trade discussions had paved the path for negotiated disarmament—a connection lost in the headlines and underappreciated by historians. Weeks before the Moscow summit, Kissinger dispatched a delegation from the USDA to meet the Soviet minister of agriculture in Crimea. They were prominent leaders in the world of agribusiness, including Secretary of Agriculture Earl Butz, Assistant Secretary Clarence Palmby, and Clifford Pulvermacher, a representative of the USDA's Export Marketing Service.[19] Prior to 1972, total US-Soviet trade amounted to little more than $200 million annually.[20] Kissinger believed that commodity trading could remake the

Soviet economy from the inside out. In private remarks to Congress, he argued that increased trade would "leaven the autarkic tendencies of the Soviet system, invite gradual association of the Soviet economy with the world economy and foster interdependence that adds an element of stability to the political equation."[21] The American delegation went to Crimea preaching the gospel of agribusiness, believing it could transform the Soviet Union as it had the United States.

Initially, the State Department and USDA officials justified the grain deal on humanitarian grounds. Since the 1930s, the West associated the Soviet Union with famine, and Soviet leadership in 1972 did nothing to alter that association, at least while negotiating terms. When the minister of agriculture, Vladimir Matskevich, welcomed a three-man agribusiness delegation to Crimea that April, he took them to see signs that pointed to a devastating shortfall in the Russian winter wheat harvest.[22] A deal to bail out Soviet harvest failures was not without precedent in 1972. Just eight years earlier, American officials had completed a similar deal, even with Cold War tensions at a peak. In 1972, fear of famine also resonated with growing American awareness about the environment and the dual pressures of agricultural development and global population growth. Ecology was a new word in American politics. In the United States, Paul Ehrlich's widely influential 1968 book, *The Population Bomb*, warned of a looming worldwide famine, the result of overpopulation and falling crop yields. Ironically, 1968 was also the year of record-busting bumper crops in rice and wheat, a phenomenon widely attributed—if not uncritically—to the experimentation of Norman Borlaug and scientists with the Rockefeller Foundation and soon dubbed the Green Revolution.[23] These record yields were said to herald a new era for Western agriculture, a capitalist "great leap forward" for agribusiness. In this spring of 1972, the American delegation went to the Soviet Union convinced of the superiority of American agriculture and simultaneously certain that the entire communist world, largely due to its backward economic ideology and the Malthusian laws of nature, teetered eternally on the brink of famine.

The historian Alec Nove deepened the perception of failed Soviet development with a 1976 article in the *New York Times Magazine* titled, "Will Russia Ever Feed Itself?" Nove blamed unpredictable weather, incomplete mechanization, and poor infrastructure for Soviet agriculture's dependency on grain imports and its vulnerability to shortages. Nove's implicit argument—that all modern countries achieve self-sufficiency in agriculture, and, by that metric, the Soviets had failed—had become a received political conceit.[24] It dated back to the first collectivization drives of the Soviet Union as well as American development programs in the United States and Mexico in the 1930s.[25] It survived into the Cold War era. As a result, the reformist premier Nikita Khrushchev made self-sufficiency a central platform of his administration. His disastrous Virgin Lands campaign, where Soviet planners brought forty million hectares of previously uncultivated land in Kazakhstan into production between 1954 and 1964, was based on this principle. Khrushchev had also looked to the United States for agricultural advice, scheduling stops in his 1959 US visit on Iowan farms and at the USDA in Beltsville, Maryland.[26] It was largely the failure of the Virgin Lands campaign in 1964—a victim of erosion, missed production quotas, and weed infestation—that led to Khrushchev's downfall. It also strengthened the legend of Soviet agricultural incompetence.

Yet by 1972, the Soviet Union had made extraordinary advances in agricultural production. The Soviets produced only forty million tons of wheat in drought-ridden 1964, bringing their average between 1966 and 1970 down to ninety million tons annually. But industrialization, and in particular the application of chemical fertilizers, monocultures, and extensive machinery, propelled Soviet agriculture to new success. In 1976, the wheat harvest hit 136 million tons, triple the 1964 yield and nearly double the longer-term average.[27] This breakthrough was largely ignored, although produced through the same methods as the highly touted Green Revolution.[28] In 1972, when the Soviet delegation came to the table to negotiate a grain deal, preventing famine was not on their minds. Nove may have decreed that the Soviets could never be self-sufficient, but the Soviet Union's decision to import vast amounts of grain was not a sign of imminent failure. Instead, it signaled an analogous transformation in communist farming, similar to changes already occurring in Western Europe and the United States. It wasn't fear of famine that was driving the push for grain imports; it was livestock production. Nove acknowledged this: "An underlying cause of the present crisis has been the attempt to improve the diet of the people. Instead of relying on bread, potatoes and cabbage, the citizens have been consuming more meat, eggs, and milk."[29]

The East, like the West, wanted more meat and used the same integrated industrial approach. An array of statistics confirms the meatification of European economies, as meat production and consumption rose steadily during the Cold War. Between 1960 and 1977, production grew between 2 and 4 percent in Eastern and Western Europe annually.[30] Due in large part to the oil crises of 1973 and 1978, which threw the economies of Western Europe and the United States into stagflationary turmoil, growth diverged on each side of the Iron Curtain. While Western Europe and its allies slowed meat production, the Eastern Bloc doubled its output, with the GDR leading the way.[31] The Soviets had made tremendous gains of their own in livestock production. The country reared 102 million head of cattle as of 1972, just 16 million fewer than the United States, and more than 71 million pigs, 7 million more than their American rivals.[32] Nove's observation—for better or for worse—was that the turn toward meat production was a sign of a small success for communist agricultural development.

GRAIN AND PIG BODIES

The industrial pig was made for détente. If rapprochement was to be achieved through agricultural deals, then an animal whose body and appetite were by design commensurable with other pig bodies and appetites could ease diplomatic tensions through trade. Like any machine or input, the industrial pig was an interchangeable cog in global meat factories. As industrial boosters standardized pig bodies for factory life, they also standardized pig diets, and thus, the fodder grown to feed animals. In this way, the industrial pig represented a profound break from its "improved" predecessors, which had feasted on local and regional sources of fodder. The industrial hogs fattened on fodder grown all over the world; feed made the industrial pig a global breed.

The shape of a pig's body is a reflection of what it eats. Whether grain based or tuber dominant, the pig's diet has exerted slow, indirect selection pressures on the shape of modern breeds, reflecting their shifting place in the broader environment. For example, in the early modern period, English hogs were left to graze on marginal common land year round to harvest the nuts (called pannage) in autumn and to forage for fungi in summer. They also ate carrion when available. These English pigs were sinewy, long-limbed, and covered in bristles, reflecting their hardscrabble existence.[33] In contrast, premodern pigs in China evolved without access to common grazing land or woods offering pannage, but rather in the backyards of peasants. Acting as the garbage disposal of the family's organic waste, Chinese hogs prospered, with short limbs, small bodies, and high fat content.[34]

In the nineteenth century, farmers categorized pig breeds not only by body type but by what they ate. In the United States, the corn belt became synonymous with its products—pork and whiskey. As the fertile soil of the Ohio Valley was first chained to the markets of New York City in the east and later

Chicago to its north and west, corn cultivation followed. The soil took to maize easily, required minimal inputs of fertilizers and labor, and produced bumper crop after bumper crop. As overproduction became a regular feature of the region, farmers hedged their maize harvests against low prices and rendered surplus corn into pork and whiskey. Historical memory of the latter now resides in the state of Kentucky's bourbon industry. A kind of terroir for hard liquor developed, built upon the association of corn, soil, and the Blue Grass state. Across the river in Cincinnati, the once famed "porkopolis" still lays claim to the invention of the modern "disassembly line." In the 1850s, farmers herded pigs along country roads and by riverboat and canal barges to feed the new packing houses of Cincinnati.[35] There, workers processed hogs moving along an overhead track, slaughtering and preserving tens of thousands of pigs a year for shipment in barrels to distant markets. As many noted at the time, corn-fed hogs had the added benefit of moving grain to market for the farmer; "fifteen or twenty bushels on four legs" went the joke.[36]

In the nineteenth century, hogs reared on corn also shared a notable trait: fat. The explosion of cheap corn in regional US markets directly shaped the direction of pig breeding, leading to the expansion of the "lard type" of commercial pig. Imported into European and then American stocks, these hogs showed their Chinese ancestry in their small, compact build, extremely short legs, and refined shape. The improved lard type also matured earlier than heirloom breeds, and most crucially, they put on lots of fat. It's no accident that the two most famous of the lard types—the Poland China and the Hampshire—were developed in the corn belt states of Ohio and Indiana, respectively.[37]

United States lard types reshaped pig breeding in Europe too. Around the turn of the century, German breeders abandoned their slower-growing domestic lard types, blaming a saturated lard market on corn-fed hogs. In response, they began to breed larger, faster-growing "bacon" types, or lean hogs.[38] The so called *Edelschwein*, or "improved," pig was created from the combination of Large White, an English meat type, and German Landrace breed. It was characterized by lighter skin, a longer trunk and limbs, and more muscle. By the twentieth century, these pigs met the growing demand for fresh pork in Europe. The *Edelschwein* was most popular among the large Junker estates of East Prussia, where grain was more readily available.[39] As a general rule, lean hogs could be found wherever available fodder consisted of dairy, peas, barley, wheat, and root crops. Breeders created new "races" such as Danish Landrace, Large Yorkshire, and Tamworth in places where corn was not king—Canada, Denmark, Ireland, and East Prussia.[40] Later, under the Third Reich's imperial, autarkic economy, Nazi breeders combined the best traits of the so-called *Veredelte Landschwein*, a breed that thrived on grain rations *and* family leftovers, with those of the *Edelschwein* to create a pig that could grow without foreign imports of grain. This new so-called *bodenständig*, or "rooted," pig embodied the ideology of *Blut und Boden*—blood and soil—since it could grow fat on crops most easily grown on German soils—potatoes and root vegetables.[41]

The industrial hog broke with this history. It was not bred to support local or regional markets. Factory conditions put new demands on the animal, which could only be met with unprecedented amounts of grain. In the Soviet Union and East Germany, standardized industrial hogs encouraged planners to look abroad for equally standardized fodder. Détente offered the Eastern Bloc their first glimpse of these new sources of feed, a fact Nixon and many other Western observers ignored in 1972. The US misinterpretation of Soviet agriculture created more favorable negotiating conditions for the representatives of Soviet *Exportkhleb*. Nixon was too busy worrying over his presidential reelection prospects. That year, his administration desperately wanted to boost turnout from his farm community base, and so they looked for new export markets for US agribusiness. Much to Nixon's chagrin, Western Europe shut them out.

By the late 1960s, Nixon and Kissinger could barely hide their disdain for European politics. The postwar economic miracles (*Wirtschaftswunder*) had continued unabated for two decades. They had rebuilt Western Europe into a self-sustaining economic powerhouse that not only competed with the United States, but actively excluded American goods from its markets.[42] For Nixon, this was a particular injustice. Only America's economic largesse and military presence had made Europe's affluence possible, he believed. Speaking in 1972, Nixon complained bitterly that "they [the Europeans] enjoy kicking the US around. Eighty-eight percent of all the European media is violently anti-US. They will cut their own throats economically to take us on politically. . . . European leaders want to 'screw' us and we want to 'screw' them in the economic area."[43] Thus Nixon's anger, coupled with a new potential market in the Soviet Union, drove his decision to send a USDA delegation to Moscow in the runup to the May negotiations.

Sensing Nixon's eagerness, the Soviets extracted a major last concession from the Americans prior to the signing of SALT I—a $750 million export trade credit allocated over three years from the Export-Import Bank.[44] The Soviets used that credit to great advantage when they arrived in New York City in July of 1972 to negotiate with representatives of North America's largest grain traders such as Continental, Cargill, and Cook Industries. While SALT dominated the headlines in 1972, it was cheap corn, cheap wheat, and cheap loans that made détente palatable to the Soviets, the United States, and the industrial pig.

MONEY, GRAIN, AND OIL: THREE "CHEAPS" AND THE 1970S CRISES

Grain represented the leading edge of commodities that began flooding into Eastern Europe in the 1970s. But others were moving in the opposite direction. This new pattern emerged clearly in early 1973. On December 30, 1972, the *Overseas Aleutian* arrived in Odessa to deliver thirty-seven thousand tons of grain. Back in the United States, a dramatic cold snap gripped the United States. Low temperatures delayed trains in the New York City subway system and Metro North Railroad, while schools across New England closed because of failing boilers.[45] Since shippers never wasted a trip by sailing empty, the *Overseas Aleutian* contracted to carry other commodities back. In the second week of January, the ocean tanker sailed out of the Black Sea bound for New York City with 244,500 barrels of No. 2 heating oil in its hold. This was the first shipment of Soviet oil to reach the United States since 1945.[46] Eleven months later, the Yom Kippur War and the subsequent OPEC oil embargo threw world energy markets into turmoil.

In the 1970s, while détente remade diplomatic relations between East and West, it was the three major economic shocks—the end of Bretton Woods in 1971, the Russian grain deal of 1972, and the oil crisis of 1973—that transformed the dimensions of global capitalism. Taken together, the confluence of new diplomatic and economic conditions generated voluminous new flows of cheap commodities—oil, grain, and money—that streamed from the Baku oil fields to GDR refineries in Schwedt and rolled through West Berlin banks all the way to Iowan corn fields, which then poured grain into ocean-going cargo ships and wound up back in feed troughs in Eberswalde. By roughly tracing the emergence of these new commodity streams, we can begin to understand how this change in global capitalism engulfed rural landscapes from the Eastern Bloc to North America over the course of the 1970s and 80s.

In 1971, Richard Nixon broke the Bretton Woods system, unmooring the dollar as the world's fixed reserve currency. This was a significant change. In 1944, the United States and its allies had created Bretton Woods to constrain the worst aspects of laissez-faire capitalism. The agreement's architects, John Maynard Keynes and Harry Dexter White, wanted a system devoted to economic stability and multinational cooperation, with states setting rules of the road for corporations and finance capital.[47] Signed by forty-six countries (including the Soviet Union, which abandoned the agreement two years later), Bretton Woods established the postwar monetary system for the West built around free trade. It established institutions like the International Monetary Fund, the General Agreement on Trade and Tariffs, and forerunners to the World Bank and the World Trade Organization. The agreement, which replaced the collapsed gold standard, aimed to maintain fixed exchange rates for currencies while also facilitating trade. The lynchpin of the system was the dollar, which the US Treasury agreed to peg to gold. The other signatories then agreed to set the value of their currencies to within 1 percent of a set value. The Bretton Woods agreement greatly constrained speculation on international currencies while also leaving national governments with the ability to control domestic monetary policy.[48]

Bretton Woods, however, required the United States' long-term commitment to global monetary stability—that is, fixed exchange rates—even if it came at a cost to its own economic competitiveness. By the 1960s, American politicians began to doubt whether this system was in the country's best interest. First, postwar economic booms from Europe to Asia changed the United States' relationship to its allies, as they could outcompete the United States with cheaper goods. Second, the Vietnam War and increased social spending created a net trade deficit in 1971, the first since the 1890s. As a result, prices for US American goods rose, even as the dollar's exchange rate remained pegged to gold. In effect, this meant that foreign countries, which needed to exchange their currencies for dollars to buy goods produced in America, could buy less. In order to push back against the artificially high price of the dollar, European countries began exchanging more of the dollars they held in reserve for gold. The so-called gold drain undermined the value of the dollar, forcing the Americans to purchase more gold to prop up the pegged dollar/gold rate. But there wasn't enough gold in the world to match the vast number of dollars that were now underpinning the world economy. Under Bretton Woods, gold continued to flow out of the American reserves. In 1968 alone, foreign governments exchanged $7 billion in gold, or 40 percent of American total reserves.[49] Over the same period, the world's financial system—largely dormant since the end of the Second World War—revived, growing to $165 billion in offshore markets in the early 1970s.[50] This newly assertive industry exacerbated the gold drain, actively speculating over changes in long and short-term interest rates, which undermined Bretton Woods' fixed exchange rates.

Increasingly, the Bretton Woods system shuddered under the expansive conditions of Western capitalism. Rather than defend this multilateral agreement, the Nixon administration decided to abandon it entirely. Without warning America's European allies, Nixon unilaterally suspended dollar-gold convertibility in August 1971, leaving the value of US currency to float. Nixon's decision irrevocably broke Bretton Woods, stopping the drain of gold from the United States and injecting a great deal of instability into financial markets. Nixon's decision also moved the foundations of US economic hegemony away from state central banks and toward commercial financial institutions.[51]

The collapse of Bretton Woods in the 1970s meant the end to fettered capitalism, especially finance capital, and thus price controls.[52] An array of actors quickly realized that they could manipulate commodity markets to win short-term profits. The Soviets, ironically, were perhaps the first to do so in the wake of the Great Grain Robbery. In the year and a half after Nixon abandoned Bretton Woods, the United States allowed the dollar to float downward against other currencies in order to boost exports. The $750 million credit granted in 1972 could now be repaid in devalued dollars and appreciating gold—conditions extremely favorable to the Soviets.[53] In short order, the Russian grain deal became a playbook for communist planners on international markets. As we will see in chapter 7, the East Germans also tried to manipulate commodity markets through large purchases and sales of grain and meat in an attempt engender a currency windfall.

Grain was not the only commodity to undergo a dramatic price adjustment in the 1970s. A similar story unfolded with oil, although unlike grain, the low price of oil subsidized the entire economies of Western Europe and the United States. The crises emerged suddenly following the Yom Kippur War of 1973 and OPEC's retaliatory oil embargo against the United States, when the price of oil rose from $3 a barrel to $11.[54] This increase was especially disruptive to economies of Western Europe and the United States, which had remade their energy regimes around crude oil since the Second

World War. In 1950, coal met 83 percent of Western European energy needs, while oil represented only 6.8 percent. By 1970, those numbers had nearly reversed, with coal falling to 29 percent and crude oil rising to 60 percent. The price of oil had also been remarkably stable. Between 1951 and 1973, the price of a barrel of oil rose from $1.93 to $2.18—an actual decline when factoring in inflation.[55] The embargo injected economic chaos into a Western capitalist system just emerging from the shocks of Bretton Woods and the Russian grain deal.

The oil crisis also created an opportunity for the Soviet Union and the Eastern Bloc. The major producers of oil were, of course, spread across the Third World. In Europe, however, the Soviets were major players in this new emerging market, supplying Scandinavia and Southern Europe with oil.[56] At first, the Soviets used their oil wealth to subsidize their satellites in Eastern Europe—the GDR being one of the greatest beneficiaries. East Germany's energy transition for the first two decades of its existence resembled that of Western Europe's, with oil imports growing by a factor of ten between 1960 and 1979. Since 90 percent of its fuel came from the Soviet Union, which priced barrels according to a five-year rolling average, the oil crisis gave the GDR a substantial competitive advantage over its Western competitors.[57] Cheap fuel encouraged GDR planners to expand the synthetic chemical industry in the cities of Schwedt and Guben with a billion deutsche marks worth of crude oil refining technology from the West by the mid-1970s.[58] Initially, the new plants were to serve as the cornerstone of new high-tech industry, but under Erich Honecker, first secretary of the SED (Social Unity Party, the communist party of East Germany) and leader of the GDR since 1971, the factories sent their refined products—gasoline, diesel fuel, heating oil, and paraffin—abroad for hard currency.[59] By the 1980s, petroleum products were 30 percent of East German exports to the West, representing the largest share by far.[60] The GDR's strategy assumed that the Soviets would always trade crude oil at below-market prices, an assumption that proved costly in the late 1970s.[61]

The oil crisis had another indirect effect on Western trade relations with the Eastern Bloc. Outside the Eastern Bloc, crude oil was priced in dollars. These internationalized dollars, known as petrodollars, had developed their own exchange markets in the first two decades of the Cold War. After 1973, the spike in the price of oil sent a flood of petrodollars to OPEC's oil-producing states—some $140 billion between 1973 and 1977 alone.[62] The OPEC nations, however, had little in the way of a financial sector, and so they invested more than half of their petrodollars in Western European and American banks. The flood of capital into Europe, however, found few investment opportunities. Western European governments were spending tax money to ward off recession and to mute the inflationary pressure on commodity prices touched off by the end of Bretton Woods and the Russian grain deal. Easy money meant fewer investment opportunities for Western banks. Flooded with petrodollars, these banks looked for new customers in new markets.

Much of this petrodollar capital went to Latin America, as well as across the Third World, to purchase cheaply produced raw materials and food.[63] Between 1972 and 1980, sovereign debt exploded across the Third World, rising from $125 billion to $800 billion in Mexico, Brazil, Argentina, and other countries. By 1980, Third World countries owed more than $750 billion to foreign investors. The vast majority of this money went toward building up domestic industries such as steel production, shipbuilding, or petrochemicals, a strategy known as import substitution.[64] Many of these countries adopted the industrial agriculture model. Unsurprisingly, cheap credit went hand in hand with cheap grain. As they had with the Soviets in 1972, the American government sent US Agency for International Development and US Feed Grains Council delegations to supposedly food insecure parts of the world, opening new routes for the global grain trade. Much like the Russian grain deal, these grain sales steadily drew the emerging Global South into a world system of industrial livestock production.[65]

Eastern Europe also participated in this process. Import substitution found enthusiastic boosters in the Soviet Union, Poland, Czechoslovakia, Romania, Hungary, and East Germany, where collective foreign debt rose from $5 billion in 1970 to $81 billion by 1981.[66] In the GDR alone, indebtedness rose from $1 billion in 1970 to $11 billion by 1980.[67] In retrospect, this borrowing would appear to have been a colossal mistake. Yet it made sense in the global economic context following the end of Bretton Woods and the oil crisis of 1973. The surge in the money supply in the middle of the decade was also driven by rock-bottom interest rates. In the West, these collective economic forces produced extreme inflation and slow economic growth, a condition that became known as stagflation. They also produced curious discrepancies in the cost of money. As Jeffrey Frieden explains, "In 1974 . . . , while consumer prices rose 12 percent, the US Treasury paid 8 percent to borrow (with 6-month securities)," a difference that expressed an effective negative interest rate for the US lending.[68] In other words, money had become nearly free. Free money was hard to turn down. As Honecker, the future leader of the GDR declared in 1973, "In today's world only fools wouldn't take up loans."[69]

As we have seen, cheap capital, grain, and oil defined the new global order of capitalism in the 1970s. They connected parts of the globe that Bretton Woods had largely ignored over the first two decades of the Cold War. The case of East Germany and its industrial pig sector shows the extent to which cheap money moved to the center of the regime's economic planning. In the short-term, free money allowed the GDR to spend on social programs, import more consumer goods for East Germans, subsidize education, and build new housing. It also dramatically reoriented the GDR economy. In 1970, 27 percent of East German exports were sent to Western Europe; by 1985 that number had ballooned to 40 percent. Western European imports grew from 22 percent of all trade to nearly 50 percent in the same period.[70] This shift in East Germany's economic orientation, however, was anything but inevitable. In the early 1970s, détente and the global economic downturn opened a breach in the Iron Curtain through which cheap capital and Western grain flowed. For a time, East Germany's decision to embrace these new resources came down to the personal biography and political priorities of its new leader, Erich Honecker.

HONECKER'S EAST GERMANY IN THE ERA OF DÉTENTE

Erich Honecker seized power in 1971. The timing was fortuitous, if not auspicious, as it placed him in control just as the tectonic plates of Cold War diplomacy and economic isolation slipped under each other. The GDR shifted too, toward a policy stance that Honecker called Unity of Economic and Social Policy.[71] Honecker believed that the legitimacy of his regime rested upon the material satisfaction of the general population. To keep the peace, Honecker sought to increase domestic consumption. He pushed policies that made material goods, such as clothing and household appliances, more available. He ordered a new wave of construction to expand the country's housing stock. He introduced or increased subsidies for education, daycare, and food. In Honecker's view, if people believed that the "radiant future" once deferred had arrived, they would appreciate the comfortable, secure societies of "real existing socialism."[72]

Announced at the Eighth Party Congress in June 1971, Honecker's Unity of Economic and Social Policy had as much to do with his past as it did with the present condition of state socialism in the GDR. As a young boy, Honecker had witnessed the hardships caused by the First World War and the German hyperinflation of the early 1920s. He came of age during the Great Depression and the rise of Nazism. As a member of the German Communist Party, he spent the entire Second World War in a concentration camp, which lent him considerable moral authority during his rise to power. Having lived through decades of economic deprivation, Honecker was convinced that socialism needed to guarantee the basic human needs of food, shelter, and work for all.

Honecker also believed that he had learned the hard lessons about the state's relationship to social unrest, having witnessed or taken part in two of East Germany's most oppressive actions in 1953 and 1961. As a member of the Politburo, Honecker had first-hand knowledge of Ulbricht's decision to crack down on the worker-led demonstrations and subsequent uprising of June 16–17, 1953. In 1961, Honecker was put in charge of the construction of the Berlin Wall. He never forgot how these two monuments to repression bookended a decade of economic austerity. Even after becoming first secretary, Honecker watched neighboring Poland struggle to contain major protests against high food prices throughout the 1970s. As a result, Honecker was determined to never let food shortages or the cost of living spark a popular protest in the GDR.[73]

Honecker decided that the Polish protests of 1970 over food price increases were the final warnings, and therefore he had to seize power from Walter Ulbricht.[74] Honecker justified his palace coup on the grounds that Ulbricht's overtures to the West for improved diplomatic and economic ties would be fatal to the GDR. In Honecker's view, Western borrowing was tantamount to socialist suicide. As we saw in the last chapter, Ulbricht had boasted to the Soviet foreign minister in 1970 that he planned to borrow as much money from the West as possible in order to spark an economic great leap forward. Equally worrying to Honecker was the issue of *Ostpolitik*, or German-German détente. In March 1970, West German chancellor Willy Brandt traveled to GDR territory for the first time since the war and met with GDR's state minister Willi Stoph in Erfurt. It was the first of a series of direct talks between the two Germanys.[75] Honecker threw *Ostpolitik* in Ulbricht's face as a "pull to the West" that had to be resisted at all costs.[76]

Once in power, Honecker announced that he was setting East Germany on a new path, albeit one that looked eerily similar to his predecessor's. In

December 1972, Honecker concluded the Basic Treaty with the Federal Republic of Germany (West Germany) and Brandt, supposedly ending years of economic (and sometimes military) hostility. The détente treaty, which allowed for trade deals and visas for foreign (mostly West German) visitors, gave Honecker room to set his economic policies in place. As a consequence, East German access to foreign capital and goods grew. Using the Unity of Economic and Social Policy as his blueprint, Honecker increased borrowing and technology imports, expecting a boost in domestic economic production and an expansion of social spending. The centerpiece of this revamped socialist economy was to be in the field of microelectronics and data processing. Between 1971 and 1980, Honecker poured billions of marks into the program, envisioning his socialist state as a leading international producer of microchips, processors, and computers. Despite rising into the top four sectors of the East German economy, his vaunted hi-tech microprocessor industry lagged far behind its international competitors. The United States, Japan, and even West Germany produced computing technology that was five to ten years ahead of East Germany's in development.[77] Unbowed, the regime continued to pour resources into the program—twenty billion East German marks and four billion West German marks between 1981 and 1988.[78] The microelectronics industry continued to absorb enormous resources while never becoming a producer of real value.

East Germany's industrial pigs told a different story. For almost two decades, the country's planners, agricultural scientists, and farmers had overseen the transformation of the agricultural sector into a potential player in global markets. As we saw in chapter 1, the regime imported new machines, expertise, and even animals from abroad to create East German factory farms. "Hog cities" such as Eberswalde, but also chicken, cattle, and sheep "cities," sprang up by the end of the decade. Yet it wasn't until the collapse of Bretton Woods in 1971 and the oil crisis of 1973 that industrial meat production took off. It was under these conditions—cheap money, cheap grain, and cheap oil—that the GDR first ventured into global agricultural markets. The new constellation of economic and diplomatic relations took the cap off the East German economy. For a brief time, there was no limit to borrowing, and thus no limit to the growth of agriculture. Engagement with global finance and détente transformed a semi-autarkic East Germany into globally enmeshed planned economy—the very thing Honecker had feared about Ulbricht's regime.

CONCLUSION

The story of the GDR during this time period is emblematic of these global transformations. At the beginning of the 1970s, cheap grain and cheap loans tantalized the East German regime with the prospect of a permanent economic breakthrough. A steady and regular supply of high-quality grains and fodder plus low-cost loans were the foundation of the East German agroecological transformation. The investments in Eberswalde and other GDR pork processors paid off quickly with a steady export product that could be sent to the Soviet Union with cheap oil flowing back in the other direction. As we will see in the next chapters, the cost of the "cheaps" rose sharply in the 1980s, throwing a wrench in Honecker's economic priorities and eventually challenging the legitimacy of his rule.

In 1972, the year of the Great Grain Robbery, these transformations still waited a decade down the road. And yet that summer heralded a new set of relations between superpowers, and shortly thereafter, their client states in their respective spheres of influence. The hard division between East and West began to dissolve in hotels, boardrooms, and cargo ships across the United States. Soon, ardent communists and true-believer capitalists discovered they had more in common than they ever believed. As Cargill vice president Walter Saunders remarked in a post-deal assessment, the Soviets "knew our grain market as well as anyone I've ever seen. They bought grain like you'd buy a used car."[79]

The Great Grain Robbery also proclaimed a new set of ecological relations, as more of the world's productive landscapes fell under the grain-oilseed-livestock complex. The "meatification" of global diets is a transformation with which we are still reckoning. It has sown the seeds of our multifaceted, uneven, and unabated climate catastrophe.[80] Yet it is history that still omits the role of the Soviet Union, the GDR, and the Eastern Bloc in spreading meatification beyond the frontiers of capitalist development. As we can see from the tale of the Great Soviet Grain Robbery, the communists knew how to operate in the cutthroat world of world commodity trading. The East German state also knew how to gamble in global markets. They played for high stakes—affordable housing, cheap plentiful food, and a low cost of living for East Germans. The chips were pigs.

CHAPTER THREE

THE SHRINKING INDUSTRIAL PIG

ONCE UPON A TIME—JULY 1982 TO BE EXACT—THERE WERE TWO NEARLY identical collective farms (LPGs), located within the southeastern border district of Löbau near Dresden. Both LPG Großhennersdorf and LPG Großschweidnitz operated as newly christened LPG Ts, or collective farms for livestock production. Rather than multipurpose operations raising grain, potatoes, and animals, their work had become radically specialized. They were emblematic of the last great land reform in the GDR, the Grüneberg Plan. Named for the country's top agricultural official, Gerhard Grüneberg, the scheme had been endorsed by the Politburo in 1976 as the standard model for all collective farms. Its central innovation was the full specialization of each LPG through the complete separation of plant and animal farming. The new economic modules would complete the transformation of the GDR "from an import land into an export land," a long cherished goal of the East German regime.[1] It was that July of 1982 that the two LPG chairmen in Löbau noticed a difference in their hog farming.

LPG Großhennersdorf encompassed four different livestock holding facilities and reared four thousand pigs, of which 410 were breeding sows. In addition, it kept 4,350 cattle on its grounds. Großschweidnitz also oversaw four thousand pigs, of which 450 were breeding sows.[2] While both collective farms held the same number of pigs and had experienced nearly the same amount of technological development, an increasing number of sows in one farm were becoming less fertile. The production figures collected by the district revealed that Großschweidnitz's sows had given birth to, on average, nearly three fewer piglets than the sows in Großhennersdorf. Wanting to know the cause, district officials sent experts to Löbau from Dresden's Institute for Veterinary Studies and the Working Group for Biotechnical Methods in Swine Science. After taking blood samples from sows on both farms, they discovered that the ailing pigs of Großschweidnitz suffered from extreme iron deficiency. The immediate cause, they determined, was the poor quality and low supply of fodder on that LPG T.[3]

A closer inspection, however, revealed that fodder was not the sole cause. Both LPG Ts had fodder contracts with neighboring LPG Ps (collectivized grain farms). Both operated under a similar tight supply of crucial mixed-grain feeds, known as concentrates. According to the district office, the difference was how each farm and its chairman went about securing, procuring, and producing that fodder. The success of each farm depended on its ability to improvise and scrounge for quality fodder on its own. The successful Großhennersdorf had secured extra potato rations from their partnering grain collective farm, guarding against a shortfall in concentrates or protein rich fodder. In addition, the manager of Großhennersdorf had gone to a neighboring dairy and acquired their surplus whey to feed to his pigs. Finally, Großhennersdorf had sown and harvested nutritionally beneficial legumes, such as lucerne (alfalfa) and clover, for sows and piglets on the small plots of land between the barns. The farmers of Großschweidnitz only had grasses and legumes to augment their fodder supply—a solution that seemed appropriate, but in reality only covered 30 percent of their nutritional needs. As birthrates fell and costs per pig rose, the farmers of Großschweidnitz struggled to meet their quotas.[4]

The story of faltering LPG Ts is illustrative of the unintended consequences of the Grüneberg Plan. While designed to boost exports through the application of industrial techniques to the farm, the plan's extreme specialization radically reduced agricultural production to a narrow band of tradable commodities. Livestock farmers, who once used their fields and pastures to supplement their animals' diets, found themselves cut off from arable land. Since the Grüneberg Plan had outsourced fodder supply to so-called LPG Ps (for crop farming), LPG Ts were left vulnerable to sudden shortages, so all chairmen had to improvise, often making things worse for themselves and their animals. The story of sow malnourishment and infertility in the district of Löbau opens a window into the bizarre reconfiguration of agriculture labor and farm life after the Grüneberg Plan.

By the mid-1970s, the GDR's political elite believed that the country's farms were in the midst of an industrial revolution. Erich Honecker's turn to the West gave the regime access to cheap credit, grain, and agricultural technology, all of which they used to build massive, vertically integrated factory farms. Mega-facilities like Eberswalde were supposed to serve as the engines of a meat exporting powerhouse, boosting the economy in the short term while unleashing a broad-based revolution in agricultural production that was to trickle down to the rest of the country. The agribusiness revolution, however, lagged on collective farms, and the main challenge was grain.

Space was a major obstacle. There was simply not enough room in the GDR to produce millions of livestock and the grain to feed them while also adequately supplying every grocery store, factory kitchen, and school cafeteria. In the meat sector, this meant that the GDR had to rely on grain imports to sustain steady expansion. The influx of cheap grain and credit at the dawn of the 1970s had seemed to offer a way out of this caloric trap. By the middle of the decade, however, those once cheap inputs were becoming increasingly expensive. Despite borrowing heavily to get the industry off the ground, the GDR had yet to produce a trade surplus by 1975. If Western credit couldn't solve the problem, East German farms would have to get bigger. They had to expand, Secretary Grüneberg reasoned, to the size of those in "the Soviet Union, but also the United States, West Germany, and other capitalist lands."[5] And his eponymous plan was the solution.

From the regime's perspective, the Grüneberg Plan was the last step in an agribusiness revolution. From the perspective of East Germany's pigs, the story was much different. This was especially true in the case of their diets and growth. Famously, pigs eat like pigs, meaning they eat anything. That doesn't mean, however, that they should. Pigs are actually quite sensitive to what they eat. Their health, and especially their growth, requires a diverse and balanced diet. We can see in the very bodies of these pigs how they were reshaped to suit the new technologies, an expensive imported diet, and the specialized production requirements prescribed by the Grüneberg Plan. In the GDR, the industrial pig's rapidly growing bodies, or worse, slow decline and malnourishment, became a visible index of industrial agriculture's success and failure.

The Grüneberg Plan, then, existed on two levels. In the political arena of the Politburo and its respective ministries, the plan satisfied major criteria of rural development ideology: it structured agriculture along industrial lines, embraced national self-sufficiency as a realizable goal, and foresaw an eventual end to Western grain imports. On the farm, however, it was a carnival of unexpected consequences as the grain and livestock LPGs grappled with the Grüneberg Plan's directives. Increasingly the success or failure of an LPG T depended on the ingenuity of the farm chairman and his ability to secure nutrient rich fodder from other LPGs and often the broader LPG community. Overlooked in their mad scramble for higher profits or increased output was a chilling metric of agricultural success right under their noses—the sick, undersized bodies of the Grüneberg Plan's industrial pigs.

CREDIT, DEBT, AND AGRICULTURAL DEVELOPMENT IN THE HONECKER ERA

Erich Honecker seemed unbothered by hypocrisy in politics. When he replaced the country's first postwar leader, Walter Ulbricht, in 1971, he had rallied his political allies to the charge that his predecessor had undertaken a dangerous "pull to the West." He pointed explicitly to Ulbricht's market-like reforms, ongoing negotiations with the FRG, and reliance on Western debt. Yet almost immediately after coming to power, he abandoned his once strongly held objections to debt and argued instead that "a substantial part of these debts to the capitalist states should be converted into long-term loans in order to ensure appropriate growth of the economy."[6] At the pivotal Eighth Party Congress in 1971, where Honecker announced his Unity of Economic and Social Policy plan, he included the expansion of agricultural production and trade as a part of his calculus.[7] Taking their orders, East German planners purchased grain and protein fodder from abroad to augment their growing livestock populations. Between 1970 and 1976, foreign food imports (including grain) rose from DM 5.7 million to over DM 11.4 million, with East German food exports rising as well, from DM 1.4 million to 4 million.[8] By 1975, Western Europe supplied the GDR with more than nine hundred thousand tons of grain, or nearly 40 percent of its total grain imports.[9] This foreign grain underwrote the expansion of the country's livestock operations, providing the necessary fodder to expand populations, as the number of cattle, chickens, and pigs rose—in the last case, from 9.6 million in 1970 to 11.5 in 1975, a 20 percent increase.[10] According to the Ministry of Agriculture (MLFN), livestock production had outpaced projections by 13 percent in 1974, an outstanding success.[11]

The regime also purchased equipment from abroad. In 1975, Honecker's government completed a deal with the West German firm Berlin-Consult to construct the slaughterhouse at Eberswalde/Britz, the third part of the massive Eberswalde complex. Berlin-Consult was a product of détente, founded in 1968 as a joint venture between the Berlin Senate and a West German agricultural investment firm.[12] In time, Berlin-Consult would build industrial slaughterhouses in Poland and the Middle East, but the 1975 deal was its largest yet, costing the GDR DM 225 million. When opened, the turnkey-ready factory processed 3,200 hogs and 560 cows per day into an assortment of meat products, the majority of which was earmarked for export abroad.[13] What set the deal apart from earlier transactions with the West was the extent to which Honecker's regime relied upon Western credit. Of the DM 225 million purchase price, 60 percent was financed through a loan from the West German Berliner Bank.[14] So long as grain and capital imports remained cheap, Honecker believed, the GDR could easily offset the costs with exports of animals and meat.

Members of the State Planning Commission (SPK) were less sure. Its chairman and the chief opponent of borrowing, Gerhard Schürer, tried to limit indebtedness, drafting a plan in 1971 as soon as Honecker came to power, only to see the first secretary change his mind.[15] In the succeeding years, growing economic uncertainty raised the SPK's fears further. In the spring of 1974, the Soviet Union informed the East German regime that they would have to introduce a series of price increases for raw material exports in reaction to rapidly rising world prices.[16] The following year, Comecon (the Council for Mutual Economic Aid) abandoned the previous commitment to fixed five year prices for trade deals. For planners dependent on consistency in their calculations, price volatility made them uneasy.

Partly out of deference and partly out of fear, the SPK refused to seriously confront the first secretary. Instead, they sought solutions elsewhere. Beginning in late 1974 and continuing into the following year, the Planning Commission engaged their comrades in the MLFN with the problem of rising costs in agriculture and the future of grain imports. From the lofty heights of a planned economy, the SPK saw the twin problems of indebtedness and grain production as a single problem of trade balances—and both were way off kilter in 1974. According to the SPK, the collective farms were costing the country more than they were earning. If agriculture, or any sector of the economy, was going to rely on foreign credit to function, it would have to follow the principles of balance of trade and export surplus. In Schürer's words,

"The take-up of further credits is only acceptable . . . with regard to economic strength, if equalized trade balances can be ensured for the next years, and beyond that, an export surplus for financing interests and services."[17] Put another way, for every quantity of something produced, there had to be an equal quantity in need by another part of the plan. For the SPK and Schürer, agriculture had grown beyond the carefully weighted balance sheets of the planned economy. It was too reliant on expensive foreign grain to maintain itself, and if it was unable to increase its output, it needed to be restrained.

In January of 1975, when it became clear the Soviets would miss their grain delivery quota by three hundred thousand tons, the SPK proposed a rebalancing of the economic plan. Rather than making up the grain shortfall with Western imports, Schürer and the SPK proposed cutting grain imports from the West further. To balance the cuts in inputs, it required an equal cut in outputs. In this case, the SPK argued to cull the cattle and pig populations further than planned and use the surplus meat—including 63,000 tons of pork it would yield—to bolster a planned export of 120,000 tons to the West.[18] By rebalancing agriculture's overall levels, the SPK believed the deal would save valuable hard currency two ways—first, by not spending on further imports, and second, by yielding a tidy profit in boosted meat exports.

The MLFN, led by Gerhard Grüneberg and Agriculture Minister Heinz Kuhrig, pushed back hard. They had long held that agriculture was the most under-funded sector of the economy, and now that Honecker's turn toward the West was benefiting rural East Germany, they saw sudden cutbacks as an affront.[19] When the SPK presented statistics that showed agriculture consuming more of its planned budget than expected, Grüneberg argued that it was a result of overproduction in grain and livestock farming, and therefore a symptom of the successes of investment.[20] Rising production numbers in animals and cereals had merely created rising costs, he reasoned. On many farms, greater investment and new cooperative relationships boosted output, but it also raised short-term expenses, such as more labor hours, more inputs like grain and fertilizer, better facilities, and highly qualified and thus highly paid workers.[21] Contrary to the SPK's views, Grüneberg believed agriculture was more than pulling its weight.

The cuts to grain imports, warned members of the MLFN, risked much more than the plan's fulfillment; they put into jeopardy the future of the countryside's social and economic stability. Ministry member Dr. Besser argued, "The proposals [of the SPK] contain fundamental changes for the development of agriculture and would have consequences for the social development of every LPG. The reduction of the productive foundations of socialism [reductions in the livestock populations with it the stagnation of related industries] is politically and economically, both internally and externally, not possible."[22] Dr. Besser maintained further that reduced output would lead to less investment and thus higher prices for farmers, putting "more distance between the development of the working class and the class of collective farmers," implying that no less than Honecker's unity of economic and social policy was at risk.[23]

Ultimately, the question of Western borrowing came down to priorities: Could the regime assure a higher standard of living, and access to bread, butter, and pork, while also maintaining its good faith and credit with foreign lenders? Or would it have to turn away from Western trade and finance and live within its means? As it turned out, the MLFN and SPK didn't have to choose. The answer lay under their feet. If they wanted to grow more food and maintain sustainable debt levels, they'd have to make soil—their most important, cheapest, and most abundant agricultural input—more productive.

THE RISE OF THE GRÜNEBERG PLAN

Over the course of 1975, the MLFN, in consultation with the SPK, worked out a major economic reform of agriculture that became known as the Grüneberg Plan. The plan remade the agricultural environment of the GDR. It expanded the scale of farm operations by increasing the overall size of fields through the removal of all human-made structures and natural obstacles. At the same time, it consolidated all livestock into fewer yet larger facilities and deployed more fertilizers and pesticides to the larger fields to augment overall output. From the perspective of the MLFN and SPK, it simplified the labor of the country's collective farms, allowing LPGs to focus on the production of a few lucrative—that is, tradable—commodities while also concentrating state investment in technology, machinery, and energy in fewer, more technologically advanced farms.

The Grüneberg Plan was the culmination of decades of development ideology. It emerged from the regime's faith in industrial development, a belief that allowed the SPK and MLFN to reconcile their previous differences. By centering the plan on industrialization, Grüneberg was declaring that agriculture could yield enough food to raise rural incomes while simultaneously provisioning the entire East German population. Trade surpluses of agricultural produce, raw materials, and fodder would also increase. As

Grüneberg himself wrote, "The conception is correct, that we must strive toward the improved supply of the population of the GDR with food, but also to do so increasingly through domestic production. It is also intended that the GDR will attempt to achieve an export surplus—meaning that more agricultural products will be exported than agrarian raw materials, fodder, and produce imported." Yet he cautioned, "How this will be accomplished is the subject of intense debate."[24] As Grüneberg and his colleagues argued over the correct strategy for agricultural development, extreme specialization offered an appealing compromise.[25]

Heinz Kuhrig presented an early version of the plan in a memo to his superior Grüneberg on April 16, 1975, that furthered the "socialist intensification of production through the transition to industrial production."[26] Yet Kuhrig also noted that crucial differences between grain farming on the one hand and livestock farming on the other complicated the transition to specialization. Industrial livestock farming would require new, expensive facilities, each specialized for a single step in the lifecycle of the animal. Under this model, new LPG Ts needed facilities for pig breeding, areas devoted to farrowing sows and their piglets, other barns that raised "weaner" pigs, and larger feedlots to fatten hogs. For this reason, Kuhrig wrote, "vertical cooperation" between these separate stages of the production process would also increase in importance.[27]

With so much labor, capital, and energy devoted to reproduction of the system of specialized livestock farming on the LPG Ts, LPG Ps would become responsible for all fodder production. In order to keep both types of farms functioning smoothly, the Grüneberg Plan would be dependent on the formation of "contracts" between crop and livestock farms and between all LPGs and vertically integrated food processing factories. This kind of rationalization depended on the "deepening of cooperative relationships" between LPGs. Cooperation, then, would supersede the constraints of tight supply through the coordination and dispersal of farm inputs like fertilizers, seed, fodder, machinery, and labor.[28]

The Grüneberg Plan broke the stalemate between the SPK and MLFN by giving each of them what they wanted. For the SPK, the plan increased efficiency in agriculture, decreased rural consumption through lower per-unit costs of inputs, and broke dependence on grain imports. For the MLFN, the plan guaranteed ongoing investment in the countryside in the form of new machinery, specialized livestock facilities, and improved living conditions for

FIG 3.1 Mural depicting the central premise of the Grüneberg Plan, on the wall of the former LPG in Gramzow, district of Angermünde, region of Neubrandenburg. The wall reads, "Soil, Crops, Animals." Photography courtesy of akg-images/ Jürgen Sorges.

rural workers in exchange for greater exports of pork and other meats to the West.[29] The Grüneberg Plan's supposed genius allowed agriculture to have its sausage and eat it too.

In practice, the Grüneberg Plan only served to make agriculture more expensive and less reliable. East Germany did not break its dependence on foreign grain, even as it sent more and more meat abroad. Furthermore, the plan compounded the problems of domestic fodder production, forcing farmers to find alternative fodder sources to feed larger herds, which in turn made pigs less healthy and the environment dirtier. Here then was the central paradox of the Grüneberg Plan: East German planners believed industrial development would create economic independence from both the Soviets and Western markets when in fact, it required a greater reliance on global markets for grain, capital, and oil. The first two were readily supplied by the West in the mid-1970s, and the last one was more reluctantly supplied by the Soviets over the same period.

East Germany's story was hardly unique. In the 1970s, the forces of cheap grain, capital, and oil were reshaping agriculture everywhere to conform to the demands of meatification. Environmental changes in the GDR found analogous forms around the world. Yet the particular shape the Grüneberg Plan took—and the environmental changes it produced—reflected the unique conditions of East Germany's geography and political regime. The basic premise that underlay the whole plan was the productive capacity of the soil. Land, as Erich Honecker was fond of saying, represented the "natural wealth" of the people and "our primary means of production."[30] At the same time, there was not much land to go around. Since the founding of the GDR in 1949, geography had placed a natural cap on agricultural growth. East Germany was a small country. In 1949, it contained approximately 6.1 million hectares of arable land (roughly the size of Ohio). By the mid-1970s, the number had slid back to 5.83 million hectares, the result of decades of industrialization and suburbanization.[31]

These limits were not insurmountable. The country's short supplies of natural resources had spurred surprising innovations elsewhere, perhaps none as impressive as in the field of synthetic chemistry. Without access to abundant raw materials like wood, bauxite, or cotton, East Germany's chemical industry produced a panoply of plastic consumer goods, albeit from the only raw (and toxic) material it had in abundance, brown coal (or lignite). From polyester dresses to prefabricated kitchens, and most infamously, the Duroplast bodies of Trabant automobiles, plastics allowed the East German regime to meet the consumer demands of its citizens without world markets or colonies.[32] In a similar way, the regime hoped its farmers could overcome the country's limited supply of land. Honecker exhorted his fellow East Germans to husband the soil through "the complex utilization of all intensive factors in agriculture." Combined with "science and technology," intensive agricultural production would create "higher yields in the fields and better performance in the feedlots."[33] The faith in technological development was belied by enormous anxiety over the scarcity of land and its supposed loss due to development. As Grüneberg himself argued in April of 1975, the plan would have to answer the question of "whether the land and the soil especially, would remain the largest producer of raw materials in the GDR."[34] At the time of the implementation of the Grüneberg Plan, the answer was not entirely clear. For the time being, the assumption that it could was enough. In the late spring of 1976, at the Ninth Party Congress, Erich Honecker officially declared the Grüneberg Plan the primary model for agriculture.[35]

PIG DIETS AND PLANNED ECONOMICS

It is hard to see the environmental history of the Grüneberg Plan in the archives. The abstract Politburo debates over development ideology and debt are an impediment to understanding actual changes on the ground. Yet behind Honecker's pronouncements and intra-ministry squabbles there lurked a story woven through the environment of the GDR with flows of capital, grain, and meat. It's a story that links the changing bodies of the industrial pig, their diets, and the global consensus over meat production.

Pigs require a careful balance of carbohydrates, fats, and proteins. From a nutritional standpoint, pigs are not especially good farm animals. Ruminants like cows and sheep use their singular digestive adaptation—the specialized rumen—to flourish on low-quality grassy fodder. The rumen contains microorganisms that break down heavy starches and cellulose bound up in grasses, releasing their proteins and vitamins. Ruminants regurgitate this food, now called cud, and re-chew it before moving this improved fodder through the rest of their stomach and digestive system.[36] In contrast to simpler, grass chomping bovids, swine possess a more complicated digestion and delicate constitution. Like us, pigs have a monogastric digestive system. Chewing the cud is beyond their ability. While beef cattle can flourish on a pure diet of roughage and grasses, the modern industrial pig does best on a diet that is nearly 95 percent carbohydrate-laden cereal grains and nutrient supplements, known as concentrates.[37] Thus, a pig's diet is most like a human one—heavy on the carbs—and yet, just like a human, they cannot subsist solely on those carbs. They require protein, minerals, and vitamins too. Farmers must provide supplementary protein feeds that contain amino acids. Without amino acids such as lysine, pigs cannot build muscle mass or repair tissue or bones.[38] Like humans, infant pigs require heavy doses of minerals and supplementary nutrients. Sow milk is loaded with vital nutrients that growing piglets need to develop the immune systems, fight infections, and become stronger. Piglets deprived of a balanced diet cannot catch up later and may face a lifetime of deficiencies and chronic disease, endangering growth and survival.[39] In other words, the new industrial pigs were machines solely bred to convert grain into meat. When these machines didn't get their prescribed fuel, they tended to break down. So, in order to have a functioning meat industry, a country needed to have grain and fodder supplements in ample supply—something the GDR struggled to achieve.

In 1976, when the regime introduced the Grüneberg Plan, East German planners believed they had found a solution to the fodder problem. According to one agronomist, "On some 40 percent of the agricultural land in the GDR fodder is exclusively grown. . . . The increase in grain production has had the greatest effect in the area of fodder availability. For example, in 1965 around 35 percent of the available grain was used as fodder, but already in 1978 around 70 percent is fed [to animals]."[40] Yet buried in these statistics lurked a central problem of agricultural reform: the GDR did not have enough land to devote to both fodder production and exports/domestic grain consumption. For example, of the 40 percent of land devoted to fodder, 23 percent was pastureland and 17 percent was farmland. Even by the planners' own admission, the bulk of available fodder was in the form of roughage—some ninety million tons, compared to only seven million tons of grain fodder and six million tons of potatoes.[41] While pigs can eat roughage, it can only make up 5 percent of their diet.[42] In this way, the pig's dietary needs forced the GDR to remain dependent on foreign imports, which between 1970 and 1984 averaged 22 percent of total fodder availability.[43]

The Grüneberg Plan went right to the heart of this problem. It attempted to mediate the divergent needs of pigs and people. As the GDR's fields got bigger, their street medians and sidewalks became sites of fodder production, and family kitchens were enlisted in the provision of fodder through recycled compost and food waste. At the same time, protein supplement imports fell. Oilseed cakes dropped as a proportion of total West German imports in the years between 1976 and 1980, from 5 percent to 2.9 percent.[44] East German planners believed that the diets of their pigs could be changed; the industrial pig proved not so flexible. Hybrid pigs were bred to withstand the automated conditions of the factory complex at Eberswalde, but on the country's newly specialized LPGs, environmental conditions were not so easily controlled. In the decade following the adoption of the Grüneberg Plan, all of these changes became manifest in the increasingly sickly bodies of the country's industrial pigs.

GRÜNEBERG COMES TO THE COUNTRYSIDE

The shift to specialized LPG Ps, which was aimed at increasing overall grain production, had the unintended consequence of limiting sources of fodder and constraining overall supply. None other than First Secretary Honecker proclaimed the importance of "intensive use" of every square inch of East

Germany's soil at the Ninth Party Congress in 1976. In the same year, Honecker reminded the Central Committee of the Socialist Unity Party (SED) that "all measures—from the large to the small—must be used for the provision of all agricultural produce."[45] While Honecker believed this would lead to an overall increase in production, in practice the result was much more complicated. On the one hand, specialization promoted monocultures on LPGs—a model industrial practice. On the other, it simplified land usage and rotations. As grain monocultures increased, the production of other rotation crops like legumes, grasses, and root vegetables decreased, putting more pressure on grain production to feed both people and pigs.

Planners recognized this potential problem early on. Writing to his superior Grüneberg in July of 1975, Heinz Kuhrig underlined the exact challenges of this plan. In order to increase grain production from 8.8 million to 11.5 million tons by 1980, Kuhrig wrote, an additional ninety thousand hectares of land would need to be cultivated annually. In the GDR, where land was already at a premium, Kuhrig knew that this increase in grain production would come at the expense of fodder cropland and pasture—around seventy thousand hectares annually. In addition, Kuhrig calculated that some twenty thousand hectares of potatoes would also have to be removed. In all, the planned increases in grain production would come at the cost of ninety thousand hectares of fodder crops each year over five years. Kuhrig recognized the acute problems of increased grain output but could only imagine a mechanical, industrial solution. He calculated that these losses could only be offset through the manufacturing of straw pellets and the processing of other forage crops from non-arable land.[46] With directives forcing farmers to plant more grain monocultures and simplify crop rotations, the LPG Ts in Löbau had to cast about for alternative fodder crops.

The pressures of Western debt pushed the regime to pursue moonshot solutions as well. In one particularly striking example, East German planners held out hopes for ongoing experimentation with radioactivity. Agricultural scientists were already hard at work strafing strains of winter and summer wheat with gamma rays. At the same time, East German scientists were also cultivating wheat in soil seeded with radioactive nitrogen in the hopes of inducing unheralded, beneficial mutations, such as high percentages of protein or miracle yields.[47] As unrealistic as these experiments in "miracle crop" breeding were, the East Germans were following an established tradition of harnessing the power of nuclear energy to agriculture. In the early days of the Cold War, American scientists at Brookhaven

Laboratories in Massachusetts planted an infamous "atomic garden"—a set of vegetable beds laid out in concentric circles around a retractable slug of radioactive cobalt. The logic of these experiments was simple: if plant breeding was based upon the creation of genetic mutations, then why not harness the power of a substance that induced mutations quickly and on a broad scale?[48] In the East German example, radiated wheat tantalized planners with the possibility of a way out of the debt/grain trap. In this way, it was expressive of a fanciful element in the Grüneberg Plan's logic, which applied not just to seeds but also to the soil.

On the country's LPG Ps, every possible square meter of arable land on collective farms had to fall under the plow. The example of the town of Lieberose illustrates the Grüneberg Plan in action, and how it reorganized agricultural space. Located in the region of Frankfurt (Oder), near the Polish border, the community of Lieberose and its 4,600 residents comprised one city and fifteen small villages. One LPG P controlled the 5,636 hectares of arable land, while two LPG Ts oversaw Lieberose's livestock, including dairy cows and hogs. Lieberose also had one state-owned pig breeding and fattening facility (ZBE Schweinemast), one fruit and vegetable collective (GPG), and two state-owned consumer goods manufacturers.

Beginning in 1975, the working committee of the community board undertook a massive survey of Lieberose's fields, farms, and forests. Its analysis revealed that of the community's 5,363 hectares, there were 150 hectares of unused land tied up in over 2,400 small parcels of land.[49] The committee then considered their soil type, their proximity to unused small roads and paths, the irrigation possibilities, and their suitability for other land uses, such as pasture, woodland, or farmland. They approached the owners of the "unused" land with offers to exchange their smallholdings for parcels in less-agriculturally productive areas of the community.[50] Here the issue of land use played a major role in redevelopment. The small parcels belonged to members of Lieberose's LPGs. Either because of their especially small size or possibly their location on marginal land, the original directors of the LPG had viewed these small parcels as uneconomical and left them for private use. Over the course of several decades, many of these parcels—over a thousand according to the survey—were abandoned as collective members fled for the West, died, or spent less time on their property. Collective property rights, while not paramount, still shaped how the farm approached the use of its land. Lieberose's survey, following the directives the Grüneberg Plan, merely determined which properties would best serve the priorities of monoculture.

The following year, the bulldozers and backhoes came to Lieberose. Removing bushes and hedgerows, they tacked numerous small parcels of land onto the perimeters of large fields, adding sixty-two hectares in the process. Next, the earthmovers set to work erasing the margins of Lieberose's outmoded fields, burying old irrigation ditches and smoothing over little-used tractor roads. The nearly six kilometers of roads and small parcels of forested land added an additional 3.75 hectares to the fields. In the meadows where farmers planted and harvested alfalfa and other grasses, the excavators dug trenches for a new network of irrigation ditches and drainage tiles at an estimated cost of 500,000 East German marks. Lieberose hoped to add an additional hundred hectares to their fields the following year.[51]

As the example of Lieberose makes clear, this period cannot be understood without paying attention to how farmland itself changed. By reserving the use of land for the LPG Ps only, the Grüneberg Plan limited not only the farmers' labor through specialization, but also what lived and grew on the farm. At Lieberose, officials pushed for the creation of "large surfaces" through the demolition of old barns and removal of farm ruins. Yet the removal of these antiquated structures was followed by the disappearance of fodder crops and animals from half of East Germany's farms. Now, fields belonged only to machinery, chemicals, and valuable grain.

Across the GDR, Grüneberg's plan reduced the diversity of crop land, biota, and labor. In the district of Kröpelin outside Rostock, a so-called Agra-industrial Association pushed this logic to a new extreme. Between 1975 and 1980, it reduced the number of crops it produced from fifteen to six, while shrinking the number of separate fields from eighty to twenty-nine. The expanded field size and simplified rotation increased the average field size from 325 hectares to 900 in the same period.[52] Everywhere the Grüneberg Plan was implemented, field size swelled, crop diversity waned, and crop rotations disappeared.

As a result, LPG Ts struggled without direct access to land. In her book *Daily Life in a Socialist Village*, Barbara Schier chronicled the life of one collective farm, called Merxleben, from its foundation until its dissolution in 1989.[53] In the mid-1970s, the collective farm, while no stranger to industrial development, tried desperately to hold off the Grüneberg Plan. The farmers at Merxleben had reasons to resist. They had recently expanded their operation, forming a cooperative in 1970 with a neighboring LPG in Bad Langensalza. Moreover, their LPG chairman had already moved the farm toward specialization, converting the collective farm into a pork-producing facility. Yet in doing so, Merxleben had managed to maintain control over its own land and fields.

In 1976, however, the district council required that Merxleben cede control of much its land, and this grain farming, to another LPG P, while restricting its own work to rearing juvenile pigs, dairy farming, and sheep.[54] Contractually, four surrounding LPG Ps were obligated to provide fodder for Merxleben, yet once state pressure dissipated, those collective farms failed to deliver the required fodder. As the director remembered, "We had to drive into the Thuringian Forest looking for fodder for animals, especially the milk cows. The problem was never money, as we couldn't find anyone willing to sell to us."[55] The livestock in Merxleben now had restricted access to feed in

FIG 3.2 Pigs rooting in a forest lot in Magdeburg, a strategy of a resourceful LPG chairman in search of additional sources of fodder. September 1973. Photograph courtesy of BArch, Bild 183-M0295-412/Manfred Siebahn.

the summer as well. Some animals were allowed to graze on the fields surrounding the barns of neighboring LPG Ps. In spite of this, nearly all of the 1,800 sows, their offspring, and an additional 13,000 pigs spent the entire year indoors, dependent on their human caretakers for their food.[56]

The Grüneberg Plan reduced the autonomy of specialized livestock farms and introduced precarity into their operations. Its focus on commercial grain production reduced the importance of fodder not intended for market, and thus took away the ability of LPG Ts to feed their own animals. If they wanted their LPG Ts to survive, LPG T chairmen, like those in Löbau and Merxleben, had to find sources of fodder elsewhere. Yet the failure of industrial fodder production to meet the demands of the LPG Ts, coupled with the directive to make all land productive, had a curious outcome—a "greening" of the industrial countryside. While East Germany's LPG Ps struggled to feed the country's pigs, the rest of the community mobilized to fill in the gap.

ALTERNATIVE SOURCES OF FODDER

Feeding industrial pigs was not easy, but their omnivory offered a solution. While grain fodder was an excellent source of calories, particularly for fattening hogs, it did not have to be the only source of a pig's calories. Villagers, with the encouragement of the regime and collective farm chairmen, turned to the grassy medians and "uneconomic" spaces in town to fill in the gaps. In 1977, a report produced for the Council of Economic Ministers recommended the cultivation of fallow spaces everywhere, including schoolyards, factory grounds, river banks, airfields, railroad beds, and even medical clinics. No space was unsuitable for cultivation.[57] Articles appeared regularly in the *Neue Deutsche Bauernzeitung* (New German farmers newspaper, or *DBZ*) highlighting examples of full, extensive production of fodder, vegetables, and fruit throughout the GDR. An article from June 25, 1982, titled "Yield from Every Square Meter of Land" recounted numerous reader reports, with accompanying pictures, of useless land returning to production.[58] In one picture, a smiling Frau Ursula Milles, from the district of Röbel, weeds beets along a strip of land beneath power lines. In another picture, a Trabant races along a tree-lined highway outside the city of Wittenberg as sheep graze lazily inside an improvised pen near the road. A small graphic at the end of the article asks the reader plaintively, "Where are the fallow or inefficient corners of land in your home city? How can we progress in the use of fragmented acreage? Do you know anyone who is cultivating similar fodder or other produce in your town?"[59]

Long before the advent of commercial livestock farming, pigs had been an essential feature in many German households. People in villages and small towns often kept a few pigs in their backyards, fattening them with leftovers. In the late 1970s, compost and table scraps returned to the diets of the GDR's hogs, although this time on an industrial scale. The city of East Berlin began regular collection of compost and table scraps in 1975.[60] The impetus here was not to green the environment but to feed the pigs. A report from the district of Berlin cited ongoing urbanization as the main cause of decline in the number of reliable "fodder surfaces" for the city's livestock. In order to feed these animals, the city organized collection drives from the city's cantinas, restaurants, and small shops, along with pickups of compost and organic garbage from city households. According to the city council, this critical fodder reserve had grown to an annual haul of thirty-six kilotons, which fed 52,200 pigs held on collective farms and another 7,000 privately reared hogs. Looking ahead, the city laid out plans for the construction of a fodder collection and preparation facility, where all scraps could be processed for the immediate dispersal into the feed of the city's animals.[61]

In many ways, garbage collection marked a return to hog-rearing methods that existed prior to the rise of the grain-oilseed-meat complex in global agriculture. The streets of antebellum New York City were overrun with garbage-feeding hogs, which belonged to the city's burgeoning lower class.[62] In the 1920s, more than a third of American cities with populations greater than one hundred thousand used hogs as a primary means of waste disposal. Worcester, Massachusetts, fed an urban herd of two thousand hogs on garbage, selling their pork for a profit of $59,000 over two years.[63] The United Kingdom created a nearly self-sufficient hog industry during the Second World War, when disruptions to the Atlantic trade cut grain imports from 8.7 million tons in 1939 to just over 1 million by 1943. While pig keepers halved the size of their herds, they refused to get rid of the animals entirely. Instead, small pig-keeper clubs organized around the collection of household waste and proceeded to maintain a robust hog population in the millions.[64] In the second half of the twentieth century, anti-trichinosis campaigns ultimately led Western governments to ban the practice.[65]

For East Germans, collection of table scraps on an industrial scale became an economic priority not only in Berlin, but all over the country. Articles in newspapers highlighted the benefits of reusing compost, while reporting on cities and farms working together to feed the country's pigs. One regular feature, titled "Fatten with Leftovers" (*Mästen mit Resten*), claimed that while nearly a million tons of table scraps had been collected, processed, and redirected into hog production in 1981, a large portion of organic material still went uncollected.[66] The article urged readers of the *DBZ* to share stories of their own efforts to capture wasted fodder. In Neubrandenburg, the eighty thousand residents divided the city into seven collection zones, gathering kitchen waste for the swine breading and fattening LPG Ihlenfeld, while simultaneously, two more LPGs collected fodder on their own. One reader reported that every six weeks, a working group for "scrap collection," made up of members of the city and district councils, met to discuss the pace of collection, emerging problems, and goals for the coming season. Articles in newspapers, word of mouth among citizens, and rewards for the most compost collected all helped to urge the Neubrandenburgers to achieve their goals. Other readers recounted stories of collecting leftover soup, bread, vegetables, and sour milk. Swine fodder collection also turned its sights on restaurants, hotels, and schools. For some, particularly those in small towns and villages, collecting fodder in this manner had been a fact of country life for generations.[67] What was new, however, was its collection and rendering into industrial hog feed on a national scale. Table scraps and industrial pigs represented the perfect blend of East Germany's agrarian past and industrial present. Unexpectedly, this union between old Germany and new took shape in the bodies of industrial pigs.

FERTILIZERS AND DANGEROUS FODDER

Fully cultivating every square meter of land brought its own problems and dangers. In order to assure increased output, the Grüneberg Plan placed particular emphasis on the use of synthetic nitrogen fertilizers in all types of cultivation—a common practice of development programs around the world. In 1976, even before the full implementation of his plan, Grüneberg outlined the parameters for a widespread application of these supercharged fertilizers. "Above all a further increase and stabilization of yield per hectare of grain is dependent on the combination of nitrate fertilizers and intensive grain types from the Soviet Union. Nitrate usage should rise to levels between 130 and 140 kg/ha, an increase over the present amount of 100kg/ha."[68] Of special note, Grüneberg wrote, was the decrease in overall area reserved for fodder production. These losses were to be offset by intensive cultivation of

FIG 3.3 Poster, "Housewives! Collect your kitchen scraps for the VEB hog feeding and slaughtering," 1972. Courtesy of akg-images.

crops like legumes, sugar beets, and other fodder crops. Just as important as increasing crop yields on arable land, Grüneberg pushed for increasing output on "green land," or meadows and pastures, through the generous use of nitrate fertilizers. He proposed increasing levels to 250 kg/ha.[69] Flooding pastures, meadows, and cropland with nitrates went hand in hand with the industrial philosophy of specialization. Yet it also did enormous damage to the environment and wound up in the bodies of East Germany's pigs.

Increasing use of nitrates on pastureland and cropland was also a self-defeating practice. While fertilizers produced a short-term increase in yield, their efficacy always decreased over time.[70] The regime knew this. The newly created Ministry for Environmental Protection produced a confidential report in 1979 detailing the extent of this decline. The report showed that whereas in 1965, 66 percent of nitrogen-based fertilizers were captured by field crops, the percentage had decreased to 54 percent by 1975 and was projected to fall to 41 percent by 1990.[71] These reports of a dramatic rise in nitrate fertilizers accompanied by a more rapid decline in their efficacy pointed toward an environment supersaturated with agrochemicals.

Nitrates also began showing up in the feed of industrial pigs. While villages and cities organized large-scale compost collection drives, industrial LPGs, like Großschweidnitz, were looking elsewhere. Harvesting large amounts of green fodder along roadsides yielded grasses and legumes often coated in nitrates. The *DBZ* reported a nitrate poisoning of several pigs in July of 1982.[72] According to the reporter, on July 4, farmers in the LPG Bierstedt, county of Salzwedel, discovered thirty dead sows in their barn. A district veterinarian determined that nitrate buildup in green fodder had poisoned the animals. According to the delivery contract with the neighboring LPG P, farmers who were supposed to test the fodder for dangerous chemicals like nitrates had not. The leader of the fodder brigade declared that the "green stuff" had shown no sign of over fertilization, but he failed to consider the environmental conditions, like freshness of the recently cut hay and its water content. The article reported that a court had recognized the failures of the LPG Ps, and a punishment would follow shortly for the farmers.[73]

Stories about nitrate poisoning and other environmental hazards were rare in local publications like *DBZ*. The Ministry of Environmental Protection, however, was well aware of the extent of nitrate poisonings, arguing in an internal 1979 report that the problem was much greater than a few dozen dead pigs every month. Analyses of some two hundred fodder samples around Potsdam, taken between 1976 and 1977, showed frightening levels of concentrated fertilizers. In 1976, over 76 percent of the samples came back showing concentration levels above the international limit of 500mg/kg level in freshly harvested fodder. The number dipped slightly in 1977 to 52 percent. Several of the samples show concentrations at ten times the limit. In each case, crucial sources of amino acids, like sunflowers, beet leaves, and watercress also carried high levels of nitrates.[74]

At high enough levels, nitrates are poisonous, but even at lower levels they inhibit important biological functions. Nitrates restrict the ability of animals to metabolize vitamin A, which in turn makes livestock vulnerable to bacterial infections like *E. coli* and illnesses like pneumonia.[75] In addition, they disrupt the ability of the thyroid to fix iodine, which severely inhibits growth in pigs. Most damning of all, the confidential report revealed the true extent of nitrate poisonings in the country's livestock. While official statistics reported the number of livestock poisonings in the double and low triple digits, the real numbers were much higher. Believing that most of these deaths were miscategorized (as caused by zoonotic diseases), the authors of the report believed that they were being underreported by a factor of "two to the tenth power."[76] So while the official number of pigs killed by nitrate poisoning in 1976 was thirty-eight, the report estimated that these numbers were actually closer to thirty-nine thousand. Similarly for cattle, the number in the same year was listed as 123, but the true number was probably closer to 126,000 animals.[77] These figures hinted at a large, widespread crisis.

POOR DIETS AND POOR MEAT

The Grüneberg Plan's shortcomings unevenly shaped the bodies of East Germany's pigs. In the country's model facilities like Eberswalde, the GP boosted fodder supplies, and its hybrid hogs flourished. The regime bestowed a preferential status on these commercial operations, largely because places like Eberswalde directly supplied West Germany and the Soviet Union with fresh pork. It was a different story on East Germany's collective farms. In its attempt to boost overall production of grain and meat, the Grüneberg Plan created a second class of farming operations on LPGs. And their pigs suffered for it. While gruesome disasters, like chemical poisonings, were prevalent on collective farms, undernourished and underweight hogs were much more common symptoms of the strained fodder supplies. By the early 1980s, the bodies of these industrial hogs reflected the strains of the Grüneberg Plan and East Germany's mounting foreign debts.

As we have seen, LPG chairmen had to increasingly seek out alternative fodder sources in the wake of the Grüneberg Plan and all its attendant problems. Strain on fodder supplies also meant that farmers had to use increasingly low-quality alternatives. Poorly stored grains, rotting compost, and partially decomposed potatoes were put in front of animals, resulting in increased lameness, infertility, and poor digestion. Substandard fodder could also make animals into alcoholics. Since farmers could not waste their feed, many fed poorly stored rations—say, wet silage that had fermented in dark grain bins—to their livestock. One East German veterinarian described how "cows started to expect an alcohol ration, as they have become addicted to the excessive slop feedings, bellowing and pulling at their chains whenever a tractor got close to the barn."[78] The same veterinarian then blamed the cows for high rates of alcoholism among rural East Germans. "Alcoholic cows create a community of addicts with their milk."[79]

Even when the fodder was properly stored, it could create problems. Such was the case of alfalfa (lucerne), which farmers increasingly planted as a substitute source of carbohydrates and crucial nutrients. After 1976, farmers turned to this legume not only for its nutritional benefits, but also because it provided a dense source of calories per hectare of land cultivated. Yet alfalfa and other fodder grasses are difficult for pigs to digest because grasses have thicker cell walls and therefore lower energy and nutrient content than grains. Furthermore, alfalfa contains chemical compounds known as saponins, which can irritate the stomach lining of pigs and reduce growth.[80] Alfalfa might work as a concentrate feed sometimes, but it cannot meet the needs of growing feeder hogs by itself. Alfalfa can make up 97 percent of a gestating sow's diet but should be less than 5 percent in the diet of fattening pigs.[81] Shrinking fodder supplies and bottlenecks, however, forced many farmers to feed alfalfa and grasses, regardless of the age group of the pig.

The overreliance on alfalfa slowed growth rates in feeder hogs. Impatient managers of LPG Ts often had to slaughter a higher percentage of their sows—pigs not raised for their pork—in order to make up the shortfall in their quotas. While perfect for reaching weight-based quotas, sows made terrible meat, and East Germans noticed. A regular complaint in the mountains of petitions received by the regime was the gristly quality of pork cutlets. For example, petitions analyzed for the city council of Berlin revealed regular discontent with "the 'overly fatty' meats and 'foreign bodies' in the sausages."[82] Gristly pork products reveal how the Grüneberg Plan worked on two levels in East Germany. Within the regime, weight-oriented quotas satisfied criteria of an abstract industrial ideal. In practice, it forced farmers to slaughter animals less-suited for consumption, many of whom were too old or contained within their flesh dangerous chemicals.

CONCLUSION

In their baggy summer suits, straw fedoras, and party-issue sunglasses, the heads of the East German state gathered at the dusty fairgrounds in the Leipzig suburb of Markkleeberg on the first weekend in June of 1979 to celebrate the thirtieth anniversary of the founding of the GDR. The leaders of the Workers' and Peasants' State had come to Agra 1979, the annual agricultural expo, to hail three decades of development. Contrasting the early threadbare years of the country's founding with the rationalized industrial landscape surrounding them at Agra, the East German leadership could revel in a clear story of progress for farmers and modernization for the country.[83]

In Hall 14 of the fairgrounds, visitors were transported to the East German countryside, circa 1949. Pictures and displays provided vivid reminders of the vast challenges the GDR faced in those early years, prelude to the centerpiece of Hall 14—a full-scale replica of a "typical" single-family farmhouse. There was the "rarely used" parlor and the adjacent "fodder kitchen," where the home and farm merged together. With "primitive" tools, farmers prepared feed for the cows and slop for the pigs, who were kept close at hand in multiuse animal pens. As one guest remarked, "One feels as if he has traveled to another world—a world that he can hardly believe lies only thirty years in the past."[84] Then stepping into Hall 15, Agra visitors returned to the glories of 1979. In Hall 15, guests were bombarded with statistics demonstrating East Germany's industrial prowess; the GDR had advanced into the ranks of the top-ten largest industrial economies; it had increased livestock production ten-fold and milk and egg production by a factor of 5.5. The dynamic East Germany of 1979 now produced the equivalent of total 1949 economic output every fifty days. And the primary reason for this success, the exhibit argued, was the Grüneberg Plan.[85]

Agra '79 was East German agriculture in miniature. It presented the Grüneberg Plan as the final piece in the country's long agricultural revolution. Yet this picture did not tell the whole truth. By 1979, the plan had failed to create self-sufficiency in food production. It led to an agricultural gigantism that increasingly ravaged the environment. Many farms had stopped rotating crops. Livestock were no longer allowed to graze fallow fields. Cheap commodity crops, with their necessary adjuncts—nitrates, pesticides, and herbicides—filled the landscape. The Grüneberg Plan also did little to alter the fundamental problem of imports and debt. Not only did it fail to ease demand for foreign grain, but it actually increased the amounts imported, despite record harvests.[86] Greater specialization, larger fields, and vertical integration had increased production, but also dependency. As one internal analysis predicted, in order to eliminate an expected import of 2.9 million tons in 1985, the GDR would have to expand grain cultivation to 73 percent of all agricultural land, displacing nearly all root vegetables, such as potatoes and sugar beets.[87] More industrial pigs meant that farmers needed more imported grain to produce more pork products, which were then exported to buy more imported grain. Greater outputs always seemed to require greater inputs.

As visitors to Markkleeberg and Agra '79 moved through the exhibition halls, they believed they were seeing a scaled replica of East German agriculture. Yet somewhere between the exhibition halls devoted to the GDR of 1949 and the Grüneberg Plan of 1979, there should have been another structure. This one would have been devoted to East Germany's new industrial pig. Display cases would have discussed not only their impressive size and enormous rates of growth, but also the chemicals and fatty tissue that riddled their bodies, the saponins that irritated their stomach linings, and the pathogens that made them sick. There of course would also have been room for the compost collection truck, the sidewalks and medians filled with grazing sheep, and the LPG T chairmen searching surrounding farms for an extra ration of whey. The industrial pig of Agra '79 was the sum of the industrial agricultural environment of the German Democratic Republic—an ideology and technology made flesh.

CHAPTER FOUR

THE MANURE CRISIS

ONE EARLY SPRING MORNING IN MAY 1982, FRIDGER PELTA SET OUT FOR A walk in the woods. A reporter for East Germany's official Communist Party agricultural newspaper, the *Neue Deutsche Bauernzeitung* (New German farmers newspaper, or *DBZ*), Pelta told his readers of his day trip through the scenic Vogtland in western Saxony. Pelta had chosen an experienced hiker to accompany him through the region's bucolic landscape of farmland, meadows, mountains, and deep woods. The Vogtland, which evoked German connotations of *Heimat*, stretched between Czechoslovakia in the east and the Federal Republic of Germany in the west, anchoring the German Democratic Republic to the central European plateau.[1] While Pelta and his companion had come to take in the beautiful vistas of Germania's folkloric home, they made a surprising discovery: sleeping giants.

> Suddenly I was startled by a cry: "I've discovered a sleeping, mossy giant."
>
> "What?" I called back. "Mossy giants only exist here in legend!"
>
> But after a quick investigation I discovered the real truth: there was no mossy giant, but rather a grassy, grown-over dunghill of imposing size. It was hard to estimate how long this creation of the Adorfer LPG [collective farm] and Mother Nature had been there. An old shepherd we encountered on the way also didn't know. But he did show us the way to another "mossy giant."[2]

Pelta and his hiking companion discovered another mountain of forgotten manure standing in the middle of a fallow field flanked by large hay stacks. From village to village they trekked, discovering giant after giant. When they arrived at a ridge in the Vogtlandic foothills, Pelta's companion, who also was an avid fisherman, spotted a pond through his binoculars.

> "It is almost definitely one of the numerous trout ponds in this area," he said excitedly.
>
> But as we came closer, I suspected otherwise: "Judging by the smell," I said, "I think we're more likely to discover swimming pigs than fish." Once there we recognized the body of water immediately—a makeshift pit of manure slurry.[3]

As Pelta and his companion fled the scene, they encountered a tractor trailer being driven, coincidentally, by the chairman of the Adorfer LPG. The chairman shrugged off their questions about the manure giants and slurry pools. He had only started his job recently. His LPG dealt with manure disposal in the correct and orderly fashion, the chairman declared. "The piles, he thought, must have been there for quite some time. Hopefully."[4]

Orderly manure stacks had long been a feature of German pig farms, but the manure giants and sewage lagoons that Pelta described were of a different kind altogether. They were the direct consequence of East Germany's all-out drive for industrial pork production. In a pivotal year for agriculture, the manure crisis reached a critical mass in 1982. Nitrates contaminated drinking water and recreational bodies of water, unleashing a public health crisis. A confidential report from the Ministry of Environment and Water Protection revealed that in 1978, more than 2.75 million East Germans, or nearly 20 percent of the population, were exposed regularly to high nitrate levels in their drinking water. These levels exceeded the internationally recognized level of 40 mg/l—a 1,200 percent increase in just ten years.[5] As a result, incidences of cancer in adults and deadly blue baby syndrome in infants increased, resulting in increased mortality in both age groups. The short-term fixes—shipments of bottled water, building new cisterns, closing and repairing contaminated wells—were expensive and ineffective. Rather than address the issue head on that year, the regime chose to hide it from the public, classifying all environmental data as confidential.[6]

Despite the regime's efforts, coverage of the manure crisis in the pages of the *DBZ* signaled the arrival of pollution in the public consciousness. Public forums like the *DBZ* were subject to censorship and state control, yet they also occasionally reflected public opinion. To Pelta, the manure giants and slurry ponds defiled the folkloric home of the German people because of the actions of a few irresponsible LPG chairmen. The critique, couched in the language of *Heimat,* was thus a safe one within the political discourse of state socialism. Every East German could sympathize with the article without crossing the delicate, invisible lines of dissent. While members of the regime squabbled behind closed doors over the causes of the problem and its degree of severity, few East Germans doubted the crisis was real. The signs were everywhere. All they had to do was go for a walk.

This chapter explores the emergence of manure pollution in the forest, fields, and waterways of the GDR, returning us to the industrial pig and to the notorious factory farm, Pig Breeding and Fattening Combine (SZMK) Eberswalde. It argues that two major crises of the 1980s—manure pollution and debt—emerged from the same quixotic project of the factory farm, and its ultimate goal, the production of cheap meat. As we have seen, in the 1970s, the regime steered the economy toward the West and toward global trade, enticed by the sudden availability of cheap grain and capital. By the end of the decade, however, those cheap inputs had become very expensive. A ballooning debt load to Western banks was swallowing up greater and greater shares of hard-won currency and trade surplus, threatening the long-term financial solvency of the country. The only answer, the regime believed, was to increase exports even further and win a larger market share.

Over the decade, grain fodder grew to consume more than 60 percent of total Western imports, with the United States, Canada, and West Germany as the chief suppliers.[7] As grain purchased with West German currency fed East German pigs, planners sent pork back in the other direction. At just one of Eberswalde's sister facilities, the hog complex at Neustadt/Orla, 80 percent of the annual production was earmarked for export to the West.[8] Eberswalde was similarly enmeshed in Western trade deals.[9] Rather than winning his country greater independence, Erich Honecker's turn toward the West had irrevocably entangled the planned economy with global financial markets and trade relations.

As debt drove factory farms to produce more meat, their pigs produced more waste. Inside the restricted concrete feeding sheds and barns, hundreds of thousands of industrial pigs gorged on whatever was set before them. The resulting waste stream was vast and toxic.[10] There were thousands of pigs per barn, and their waste was collected in slurry pits and lagoons. Anaerobic microorganisms and pathogens thrived there, reproducing, mutating, and releasing noxious chemicals. Storage tanks were periodically emptied, and treated waste was returned to fields to be plowed under or dumped into the sewer system to be carried to the Baltic Sea.

The problem was even more severe on the collective farms. A decade after the implementation of the Grüneberg Plan in 1976, they had become a threat to public health and to the state economy. While planners were at least aware of the dangers from concentrating so much pig waste around factory farms like Eberswalde, they overlooked the danger posed by the millions of industrial pigs spread out across thousands of collective LPG Ts. Largely due to their incomplete development, these LPG Ts were still treated like traditional holdings and encouraged to use traditional methods of manure handling. Overwhelmed by the toxic substance and unable to resell it to partnering LPG Ps, farmers were forced to truck it to marginal land and abandon it, thus creating the breeding grounds for Pelta's mysterious mossy giants. As industrial scale livestock farming began to decline in the late 1980s, the tide of animal waste receded; the memory and the anger of ordinary East Germans toward a government seemingly indifferent to their fate, however, did not.

THE SLIPPERY NATURE OF MANURE

The problem bedeviling Eberswalde's engineers was the slippery nature of industrially produced manure. Rather than an inert mass, industrial manure is a complex chemical and biological soup of water, nutrients, inorganic chemicals, organic matter, and microorganisms. It is not a uniform substance but a mixture of dung—the hot mass voided by the animal—and other organic matter like straw, water, and microorganisms that decomposes at different rates depending on environmental conditions, such as temperature, the presence or absence of oxygen, mechanical agitation, and the chemical makeup of the manure.[11] Animal manure can be divided into three basic types: solid manure, semisolid, and liquid manure. Solid manure, the easiest to manage, generally with shovels or pitchforks, is often combined with hay or other plant matter. This breaks the material into smaller particles and opens airways into the mass that allow it to decompose more readily into reusable topsoil or spreadable fertilizer. Semisolid manure, containing between 5 and 15 percent solids, is difficult to treat. It is too thick to pump and too thin to shovel. Liquid manure contains more than 95 percent water and less than 5 percent solids, and it can be flushed or pumped from one location to the next, like gray water.[12]

In the GDR, liquid manure proved the most problematic. Traditionally, the best way to treat liquid and semisolid manure was to prevent its creation in the first place. A farmer had two methods. He could change the animal's diet, supplementing grains with rough plant matter to increase the amount of solid manure voided. Or the farmer could bed the animal on straw, which would absorb semisolid and liquid manure and make handling easier.[13] The rise of the factory farm, however, made these approaches impossible, increasing both the amount of waste produced and its toxicity.

The problem lay in the function and design of industrial facilities. As factory farms accelerated the meatification of global diets, they promoted the concentration of animals while simultaneously reducing livestock-keeping into a narrow series of inputs and outputs—cheap grain and animal bodies entered in one side, meat and animal waste exited on the other. Due to this streamlining of production, factory farmers could not afford to include roughage to create better manure, as the practice slowed weight gain. Furthermore, the concentration of thousands of animals increased the sheer volume of liquid manure produced. Industrial-scaled waste required an industrial-scaled solution. Rather than using straw to absorb pig manure, engineers designed slatted concrete flooring, which acted as a sieve, separating solid wastes from liquid. The remaining watery mixture ran underneath the stalls into long drainage canals that led outside the barn. Workers cleared the solid waste from the pens before lifting the grated subflooring and flushing the remaining manure with hoses and brooms from the concrete canals.

Once outside, liquid manure flowed into separation tanks and sedimentation lagoons, where dangerous concentrations of pathogens and parasites, including salmonella, giardia, and cryptosporidium, built up. The byproducts of this pathogenic brew were harmful. While hog manure can decompose in the presence of oxygen, most industrially produced manure was left to decompose anaerobically, largely because of the enormous volume. Without oxygen, a host of anaerobic microorganisms in these pits feasted on the organic matter in suspension, producing noxious gases such as carbon dioxide, methane, ammonia, and hydrogen sulfide. Besides the stench, all these gases in high concentrations could be harmful to humans and livestock. For example, methane is highly explosive, while hydrogen sulfide, in only a few parts per million, can numb the sense of smell, effectively

FIG 4.1 A collective farmer rests after washing out the central drainage channel in an animal shed. November 8, 1990. Photograph courtesy of BArch, Bild 183-1990-1108-012/Ralf Hirschberger.

preventing detection before reaching deadly concentrations.[14] Thousands of confined pigs threw high levels of dust, airborne bacteria, and other infectious agents into the air. Livestock workers in these facilities were known to suffer from higher incidences of bronchitis, atopic asthma, and acute organic dust toxic syndrome.[15] In short, the industrial concentration of livestock turned manure from a traditional organic fertilizer into a highly toxic substance.

Even before crews broke ground at Eberswalde, the Ministry of Agriculture (MLFN) and local management knew about the potential threat of toxic manure. Since industrial pigs required seven to eight units of fodder to create one unit of muscle mass, this meant a lot of calories were lost as dung.[16] In addition, farm chairmen had to maintain a constant turnover of their hog population. On average, two generations of piglets born to each sow per year would be needed to assure ongoing production. Rapid growth rates, when combined with inefficient feed conversion and constant population turnover, created an unprecedented amount of waste. Conservative estimates calculate that a single pig produces four to five times the amount of waste of a person.[17] For East Germany, with an average hog population during the 1970s of between ten and thirteen million hogs, it was like adding the excrement of between 44 and 65 million people to a country of just over 16 million people.

If industrial livestock production was to become the dominant model for agricultural development in the GDR, a solution to the manure problem was needed quickly. From 1967 to 1976, engineers and scientists at Eberswalde experimented with new techniques for manure handling, looking to develop a model that could be expanded to collective farms around the country. Needing a technology was one thing; finding a practical one was another matter. By 1975, with the adoption of the Grüneberg Plan looming, members of the MLFN demanded answers to the manure problem from Eberswalde. There were none forthcoming.

IN SEARCH OF A SOLUTION

In the beginning, SZMK Eberswalde had no specific plan to counter the dangers of industrial manure disposal. The Yugoslavian firm Emona had overseen the design and construction of Eberswalde in 1967, but its engineers had not solved the manure disposal problem. As a result, the managers at Eberswalde developed their newly imported technology in a haphazard manner. They started slowly, with a relatively small herd of twelve kilotons (roughly thirty-five thousand pigs) in 1969.[18] The agricultural specialists and farm managers believed that effective manure handling systems would control the nature of manure decomposition. First, proper handling would reduce the toxicity of manure and lessen the danger that the manure posed to workers, livestock, and the surrounding population. Second, "processed" manure could then be reused as fertilizer for field crops, root vegetables, forestry management, and even as a protein supplement for animal fodder.

The SZMK engineers focused their efforts on oxygenating the pits. They worked first on increasing the surface area of the slurry lagoons, believing it would expose more of the liquid manure to air, hopefully breaking up or lessening dangerous anaerobic conditions.[19] If successful, the new oxygenation technology would give Eberswalde another valuable export product, "organic fertilizers" made from toxic waste. In 1969, engineers built a scaled-up sedimentation lagoon system based on the inchoate Yugoslavian design that covered an enormous twelve hectares. The design created two adjacent lagoons, the second slightly lower than the first. As the slurry poured into the upper lagoon from the pig shed, gravity pulled the denser, non-dissolved substances to the bottom, while the less dense liquids collected at the top, the spill-over then draining through a pipe into the lower second lagoon. Thus, solids would settle out in the upper lagoon while liquid manure and waste water filled the pond below. The liquid manure was then pumped either directly into an irrigation system or into spray tanks, treating around 1,450 hectares of land in the process. The solids were to dry out in the upper lagoon.[20]

The system never functioned as intended. First, sedimentation alone could not effectively separate the liquids from the semisolids, which continually clogged pipes and hoses. Second, the liquid manure still retained too much of the solid matter, pathogens, and unused nutrients. The lower lagoon silted up, decreasing storage capacity. Worse, the treated liquid manure not only failed to fertilize the soil; it leached nutrients from the fields it was supposed to enrich. Third, the bio-sludge in the first lagoon refused to dry under the cloudy skies of the GDR. Instead, it threw off a noxious odor that could be detected for kilometers.[21]

An unhappy Politburo ordered more research into manure handling technologies in 1972, establishing a permanent working group at SZMK Eberswalde. Drawing on experts from East German institutions and industries, the working group included members from the Berlin branch of the Association of People's Facilities for Industrial Animal Production, the Chemical Producers Combine at Stassfurst (CAS), and the Academy of Agriculture, as well as representatives of Institute for Manure Research from Potsdam (IDF). Responding quickly to their mandate, the working group, which was led primarily by the CAS and the IDF, published its first recommendation in January 1973. They followed that with a series of procedures for a scientific/technical-oriented pilot program for a two-level microbiological processing of manure. The pilot built on the earlier sedimentation design but proposed to increase oxygen exposure in the manure either through forced aeration or increased surface exposure. The Politburo endorsed the plan two months later, effectively announcing new guidelines to livestock farmers around the country. While the working group tackled liquid manure decomposition, the Academy of Agriculture noted in a report that Eberswalde had not yet developed a solution to the problem of overflow.[22]

The academy's report also highlighted a conceptual problem; while the manure scientists and engineers wrestled with the technology to change the makeup of the manure, they were far less concerned by the quantity. This attitude was a holdover from traditional husbandry practices in German agriculture. Before the rise of chemical fertilizers, advanced husbandry practices relied upon livestock manure to return nutrients to the field. Far from toxic, manure produced from high quality feed had many beneficial characteristics, including the return of trace elements from important soil nutrients such as phosphorous, nitrogen, and potassium.[23]

These nutrients were expensive to produce or to import. Some agricultural scientists believed that treating manure industrially would be a more efficient way to capture valuable proteins like lysine, which could be reincorporated into pig feed. From its very first days, manure research at Eberswalde focused on trace-element recapture, only to encounter difficulties such as varying levels of available protein in treated manure and the high costs of experimentation.[24] Despite these setbacks, research into creating manure "fodder" continued well into the 1980s. The East Germans even began exporting the substance as "organic fertilizer." The complex at Eberswalde developed Western partnerships with Friedrich Wilhelm University in Bonn and with several companies in Duisburg to experiment with and expand the market for manure-derived protein.[25] Most research, however, focused on the problem of disposal, yet few managers stopped to consider whether industrially produced manure should be returned to the soil at all, or if reapplied, in what quantities or concentrations.

As we have seen, 1975 was a turning point for East German agriculture. In the same year that planners worked out the principles and guidelines of the Grüneberg Plan, the MLFN looked anxiously to Eberswalde, hoping that a breakthrough in manure treatment technology could soon be disseminated to thousands of collective farms. As they were well aware, the implementation of industrial hog farming was risky. Two years earlier, the newly established Ministry for Environmental Protection (MfUS) warned the regime about the dangers of agricultural pollution. In a resolution titled the Water Management and Manure Law of 1973, the MfUS stated that increased waste was an inevitable consequence of the Politburo's economic policies as laid out during the 1972 Eighth Party Congress. The Ministry argued that the platform, which called for a "further intensification and ongoing conversion to industrial production of livestock," required a scientific approach to monitor animal-produced waste.[26] The resolution did not argue for limiting the new industrial agriculture program, but pointed out that since "intensification" would create more of everything (meat *and* manure), the GDR needed ways to assure that animal waste was safely disposed of or reapplied as fertilizer.

The MLFN feared that without a breakthrough technology, specialized LPG Ts would drown the country in manure. A detailed report of manure pollution near the city of Rostock was a nightmare brought to life. At Trinwillershagen, farmers running a recently consolidated and expanded livestock LPG discovered that the manure from their ten thousand pigs and two thousand cattle had leached into the surrounding watershed. Applied to Trinwillershagen's fields, the treated manure was reaching the Saaler River and, after pouring down through a renowned natural wonder, the chain of lagoons known as the Darß-Zingst, was flowing into the Baltic Sea. The report said that the runoff had already seriously damaged the natural lagoon system and violated an international conservation treaty to protect the Baltic.[27]

The MLFN was not amused. Frustrated by the crawling pace of innovation at Eberswalde and slow-moving disasters like Trinwillershagen, in 1976, Bruno Kiesler, the Politburo member overseeing the transition to full specialization, took the managers at Eberswalde to task. Kiesler criticized the Eberswalde managers for their failure, despite years of investment and research, to develop a practical system for the "productive reuse of manure." He railed against further "flights of fancy" from scientists and functionaries, demanding instead immediate action. No longer willing to tolerate failure, Kiesler tasked the state's minister of agriculture, Heinz Kuhrig, with the oversight of manure research.[28] Kuhrig ordered SZMK Eberswalde to begin fulltime trials on two rival technologies. One was a curved-screen press, a machine that essentially ran liquid manure through a machine, wringing water from the solids. The other process created a large decanter. After trials, ministers Kuhrig and Kiesler would decide which technology was the best.

Under growing pressure, Eberswalde's manure processing working group told Gerhard Grüneberg in early 1976 that their experiments with the new curved-sieve press and a so-called two-step, microbiological preparation had indeed converted toxic waste into useful manure. Properly operated, their new system could treat manure without generating excessive smells, heavy sedimentation in the lagoons, and disease threats—or so they claimed. The new system did not clog irrigation lines and it reduced nitrate concentrations by 50 percent.[29] The new system quickly proved too good to be true.

Throughout the period of experimentation, technicians (and planners) remained committed to the vision of rendering industrial manure reusable. The Eberswalders identified a farm for experimental treatment in the village of Lichterfelde. Using a combination of tractors, high-volume pumps, and spray irrigation, they began applying treated manure to the 1,600 hectares of land in 1975.[30] The hopeful trial soon turned to disaster. The fields quickly became supersaturated with nitrates, leaving behind visible manure ponds.[31] A 1976 study of local water tables and hydrology indicated drinking water contamination. Worse, the pollution plume was spilling into the neighboring state forest of Schorfheide, which Erich Honecker held as a prized game reserve. The technicians immediately stopped treatment on the four hundred hectares of Lichterfelde land.[32]

Limiting the scale of the experiment, however, created new problems. Eberswalde's exploding pig population desperately needed another location for their estimated 1.3 million cubic meters of manure produced by the industrial pigs annually.[33] In response, managers drew up emergency plans for the short term while sketching out a long-term solution. Initially they found a three-hundred-hectare field where they would dump manure. The plan was to return later to "sanitize" this plot of supposedly nitrate-tolerant land. Long term, the Eberswalde managers would expand the irrigation/fertilization area to the east and south of Eberswalde while diluting the toxic liquid manure with gray water from the sewers of Berlin.[34] What had started as a demonstration of the scientific progress made at Eberswalde ended in a fiasco.

Despite the horrific 1976 disaster, the managers had made no real scientific or engineering breakthroughs. Instead, the planners pushed ahead with existing technologies but on a larger scale. For example, Treatment Zone I, as Lichterfelde later became known, was not abandoned. It was simply connected to another, much larger Treatment Zone II approximately 6,100 hectares between Britz, Golzow, and Werbellin. The treatment areas were connected to farms through central sewage lines from manure pumping stations at Eberswalde. There it met another sewage line of gray water running from Berlin. This was a mega-engineering solution to an ecological limitation. In the treatment zones, irrigation lines drew from the main sewage line. They fed rolling applicators, hydraulic giants that stretched three hundred meters from end to end. With model names like the Frigate, the spray applicators were Lego-like mechanical wonders of pipes, wheels, and spray arms. For the applicators to function well, all obstacles had to be removed from the fields. Even today, the technology has left its mark in places like Britz, where long, narrow hedgerows mark the tracks where the massive manure spray applicators once rolled.[35] This obsession with technological solutions privileged the use of machinery while often overlooking critical ecological processes still at work in the farm environment.

Eberswalde's managers learned from the early disasters at Lichterfelde that the best way to limit immediate adverse effects was to spray the "organic fertilizer" in thin, even applications. The concentrated effluent from factory farms like Eberswalde, however, made safe application impossible. The industrial production of meat required significantly less land than previous methods, but the disposal of (suddenly) toxic manure required far more land than ever—thousands, if not tens of thousands, of hectares more. Yet for Politburo members like Kiesler and Kuhrig, the "quality" of the manure was all that mattered—thus their endorsement of the bowed-sieve press and decanter technologies. The issue of quantity, as the experiment at Lichterfelde showed, remained stubbornly disconnected from the issue of total area. And so in 1976, the MLFN, along with the entire Politburo, formally endorsed the Grüneberg Plan. Integral to the Eberswalde model for pork production was the unresolved problem of waste disposal. Before 1976, the manure pollution problem was confined to a few isolated facilities like Eberswalde. When the Eberswalde model was exported to the country's collective farms, the Grüneberg Plan ensured that nitrate pollution would become a national crisis.

EBERSWALDE MOVES TO THE COLLECTIVE FARM

In January 1976, the MLFN began preparing the GDR's more than three thousand collective farms for the hyper-specialization of the Grüneberg Plan. The agricultural reform would require new technical skills, especially in manure treatment. Planners envisioned the creation of "modern" farmers who, through the use of advanced technologies and machinery, would operate their LPG Ts like mini-Eberswalders. Unfortunately, the solutions these modern farmers would apply remained in an experimental phase that winter. In lieu of new technology, the MLFN had to come up with an alternative approach. Without curved-sieve presses or pumping stations, the regime turned to the agricultural traditions of the past for answers.

A farm is an engineered space. Its organization, from the crops it grows to the location and purpose of its buildings, transposes a particular vision and set of demands on the landscape. Every farm is the product not only of this engineered order, but also the millions of organisms and non-human actors that inhabit the same space. The bacteria and invertebrates in the soil, the hydrology of the fields, and the dietary needs of livestock as well as the scavenging wild animals and pests all have profound impacts on the management of the farm. Before the industrialization of agriculture, this engineered vision relied upon an interdependent relationship between crops and animals, as each provided the other with crucial sources of energy and nutrients.[36] The system, which reached its apogee in the nineteenth century, was known as high farming.[37]

The interdependent relationship between crops and animals imposed a circular logic on farm life. Without modern heavy machinery for planting, harvesting, or transport, nineteenth-century farmers relied on symbiotic relationships between plants and animals to move energy around the farm and improve soil. Rotation, in this sense, incorporated not only crops, but also root vegetables, grasses, livestock, and even trees. For example, farmers regularly turned livestock loose on recently harvested fields or onto parcels planted with fodder crops which the animals then self-harvested. Grazing livestock also fertilized the same fields with fresh dung.

In addition to crop rotation, the interchanging of leys (temporary pastures) was another way to restore soil fertility. Planting grasses, usually nitrogen-fixing clover and alfalfa, could serve two purposes, restoring soil and fattening livestock. Other fodder plants helped make soil productive again. Root crops such as potatoes, sugar beets, and carrots cleaned the soil of pathogen and pest build-up; they also broke up compacted soil with their deep roots. Pigs in particular enjoyed feasting on root crops, and, moreover, they did much of the digging themselves. Within certain limits, these practices allowed farmers to keep livestock year-round with animals feeding on summer pastures and winter root crops. Farmers augmented all of these restorative practices through the regular return of animal fertilizers to the soil. Here is the crucial difference, articulated by an old farmer adage: while high farming rotated the farm through the landscape, industrial farming attempted to rotate the landscape through the farm.[38]

The Grüneberg Plan's great innovation—the separation of plant and animal production—took modern industrial farming to a logical extreme, inaugurating an entirely new regime for the East German farm ecosystem. Heavy machinery, extensive irrigation, and agrochemicals allowed collective farmers to abandon the restorative practices of high farming and instead to increase the total area cultivated by a single farm. The size of specialized LPG Ps expanded dramatically after 1976, reaching an average, 4,130 hectares, which was more than fifteen times the size of the average collective farm at the end of the final round of collectivization in 1960.[39] The degree of livestock concentration on the new LPG Ts was unprecedented and should have made planners cautious. Instead, planners seized on it as the only logical outcome of three decades of rural development.

Throughout 1976, the MLFN held retreats and seminars for LPG chairmen to prepare them for the transition. The first meeting held in Brandenburg in January drew prominent LPG chairmen from all over the region (Bezirk) to discuss "improvement in the care of the soil with organic substance and the use of all reserves for increased fertility."[40] The chairmen developed a set of guidelines titled "First Experiences in the Production and Use of Compost."[41] The opening section listed additives and industrial byproducts that might be used to turn industrial manure into compost. To build quality humus, the top and most vital layer of soil, the chairmen recommended byproducts such as gray water, slaughtered animal offal, feathers, spoiled fruit, lignite, and even coffee grounds.[42] Once they mastered the new techniques of manure handling, the LPG T chairmen were assured that their processed manure would become a valuable byproduct for sale to neighboring LPG Ps. This steady supply would insure that all deals would eventually be formalized through "contracts."

The chairmen at these early seminars still regarded manure in much the same way that their farming grandfathers had—as high quality, nutrient-rich organic matter. But industrialization—and, in the East German case, the separation of plant and animal production—had altered organic inputs available to LPG T farmers. They would no longer have direct access to spreading fields, to straw and to other plant matter that were essential to traditional manure handling. In addition, the chairmen failed to realize that the manure produced through specialized production would be of a different quality and in far greater quantity that anyone had ever addressed through traditional techniques. After 1976, as the specialized LPG Ts came into full production, their chairmen soon realized that discussions on *how* to treat manure were beside the point. The issue soon became *where* to put all this rapidly accumulating, dangerous pig manure.

MANURE CONTRACTS AND MUCK MOUNTAINS

The manure crisis played out under the East German concept of "cooperation." To bridge the gap between high farming and industrial agriculture, the two kinds of specialized LPGs were to coordinate everything from production goals to raw materials. They were to cooperate on the sharing of labor, machinery, and, most important, the of delivery of fodder and organic fertilizers. Manure, however, was one commodity that nobody wanted. It was the most difficult to use and the most expensive to move and to store. It was also the largest product that newly specialized LPG Ts had to offer. Cooperation over manure removal and storage was critical for LPG Ts that were short on land. In order to enforce cooperation, the MLFN promoted the creation of contracts for manure handling. In theory, a specialized LPG P would determine the farm's need for fertilizer and then work out a year-long order for fertilizer with an LPG T. The contract also specified the date of manure delivery and removal, as well as who would pay the costs of storage, transportation, and redeployment in the fields.[43] Once the two collective farms came to an agreement on price (either compensated in cash or paid out in fodder), a regional body of the MLFN ratified the contract.

Transportation and storage were not small matters. Both were expensive and difficult to do well. Treatment required special tools, complex machinery, and fuel-hungry vehicles. As the promised new technologies failed to appear, the specialized LPG Ts realized that they lacked the physical infrastructure to safely handle what was clearly toxic waste. One report estimated that only two million of the country's twelve million pigs were being raised in "modern" facilities with the supposedly advanced manure processing technologies developed in 1975. The remaining LPG Ts had built their own storage tanks without proper plans, procedures, or regulations. Most of these manure tanks could only store up to a few days of waste instead of a more practical sixty to ninety days' worth.[44] For LPG Ts, which had lost direct access to pasture and cropland, manure storage and transportation became especially urgent. Without space for disposal or a new technological solution, LPG T farmers were trapped with an increasing burden of deadly toxic waste.

On paper, LPG Ps and Ts reached a compromise solution through the creation of *Zwischenlagern* (ZL) or temporary storage locations. Farmers had first used ZL, an older practice of high farming, to store harvested hay, root vegetables, and crops at the edge of large fields before moving them on to markets or to the barn. In its East German revival, a ZL could be as technologically advanced as a clay-lined slurry lagoon. All too often, the ZL was a makeshift shallow trench covered with a tarpaulin. The ambiguity of the term—particularly the implication that it was an actual structure—allowed farmers to get away with an ecological catastrophe. "Muck mounds" sprang up in marginal or fallow fields or on wasteland.[45]

The ZL became a safety valve for the inelastic terms of "cooperation" contracts and the extreme specialization of the new agricultural regime. The ZL made the waste problem seemingly disappear without seriously jeopardizing farm production goals. Yet the problem remained, even under tarps. The change of seasons presented significant challenges. Fertilization was supposed to follow a strict agricultural calendar that accounted for changes in weather and soil conditions. *Zwischenlagern* ostensibly freed collective farmers from this calendar. Chairmen of LPG Ts could keep their barns and feedlots cleaned out year-round. On LPG Ps, farmers no longer had to reapply the manure immediately but could wait until conditions or workloads were right. The ZLs opened a giant loophole in manure regulations. As neither livestock nor grain farms had any use for increasingly toxic waste, they dumped it wherever they thought it wouldn't be noticed. Thus, the mossy giants sprang up on waste ground, fallow fields, or in the woods where hikers and reporters like Fridger Pelta found them.

As the mossy giants became an open secret, the LPG Ts themselves became an increasingly contentious issue in the pages of the *Neue Deutsche Bauernzeitung*. Journalists like Pelta had raised manure disposal to the top of public discussion. Each spring between 1979 and 1982, the *DBZ* ran a section titled "Is the Soil Getting What It Needs?" The section featured letters and photos on best practices for manure handling, but they also reflected the growing public anger over manure pollution. Pelta became a leading critic. In articles he wrote or in the *DBZ* issues he edited, Pelta drew attention to three primary issues in manure handling: seasonality, storage, and cost.

In his reporting, Pelta highlighted farmers who were handling manure correctly by using the pre-industrial method. There was a right way and a wrong way to stack manure. Farmers raked and shoveled the mix of solid, semisolid dung, and straw bedding from the animal stalls and hauled it to the barnyard. By the 1970s, the more successful and favored LPG farmers had modified small tractors for bulldozing organic matter, yet most East German farmers still worked with shovel, pitchfork, and wheelbarrow to transform this shapeless muck into precise six- to ten-foot-high by thirty-foot-long rectangles. A well-stacked muck mound yielded valuable manure for grain farmers, but it was also the hallmark of a skilled farmer. According to dairyman Reinhold Hagenorn, "As my father always said—who was also a dairyman—the manure pile is the embodiment of a good dairy farmer."[46]

Hildegard Sens, a dairy farmer in the LPG Leitzkau in the district of Zerbst, called such work "a matter of honor." Writing to the *DBZ* in 1979, she noted that "it hurts to see, again and again, manure tossed carelessly from the barn and left strewn about for all to see."[47] In her traditional barn, she and a colleague, Heinrich Schnelle, cared for fifty cows. "For us, to carefully stack the manure in the yard is honorable work. Sure the work is not easy, but the praise of [cooperative] grain farmers always urged us to continue such work. For their part, the grain farmers strive to deploy our fertilizer in good time."[48] Frau Sens described a type of agriculture that was on its way out in the post-Grüneberg landscape. Not only had barns of fifty dairy cow become exceedingly rare, but so would the necessity of organic manure in agriculture.

Weather also posed new challenges for LPG Ts. During the "catastrophic winter" of 1978–79, extreme cold and record snows over the New Year brought economic activity to a halt. Coal trains stopped, and vast sections of the country went without power for weeks. Manure, heaped in barnyards or trapped in sewage canals, froze. Without electricity to heat their barns, the farmers watched cold kill their pigs. In the district of Lübz, in the northern region of Schwerin, village correspondent Gottfried Brouwers reported in early March 1979 on a farm flooded with manure. "In this dairy manure has become a major source of concern. It has already overflowed the storage tanks and comes up through the slatted-concrete flooring."[49]

In such emergencies, many livestock farmers were forced to dump manure in the nearest field, no matter the weather. Writing from the district of Salzwedel, village correspondent Bruno Loeding said winter made everything worse. "Where should manure and liquid slurry be deposited? This question confronts us farmers regularly, but even more so in winter. In the past weeks, liquid manure was driven out into the snow out of emergency reasons—a practice which is particularly aggravating when so many fields call out for organic nutrients."[50]

Pouring liquid manure onto snow-covered fields was ecological vandalism, concentrating a plume of toxic liquid slurry in a small area. It was also wasteful. The snowmelt would carry the waste farther as surface runoff. A MfUS internal report singled out this overconcentration of "organic fertilizer" as a major source of pollution. The amounts of manure being reapplied as fertilizer, the report said, was often disproportionate to the area set aside for treatment. The treatment areas were often too close to one another, "overwhelming the reproductive and cleaning capacity of the soil."[51] The problem worsened in the winter; "Since only half of the necessary storage capacity is available, dung has to be brought out to the fields during the winter months."[52] Many years later, one LPG member recounted how manure dumped in the snow over-fertilized fields and damaged crops: "We sought out a field and spread liquid manure on it during the entire winter and plowed it in early spring. Then corn was planted there. The corn that grew there became dark green and was so loaded with nitrite that it would affect the cattle."[53]

Even in the warmer months, LPG Ps had little use for animal-based manures. Increased specialization in field crops had reduced the scale of crop rotation, and with it the number of opportunities to spread manure. Noticing the shift, some LPG T chairmen advocated for a return to more regular rotations and catch-cropping, which would keep manure form washing out by limiting fallow periods. "On many of our fields, grain is often planted for four or five years in a row. Since the harvests occur so regularly, there is no room for intercropping. We should try planting a green manure rotation (a mixture grasses and legumes) followed by potatoes. This would increase the opportunities to fertilize with our [animal] manure."[54] Not only did organic fertilizers lose out to increasingly simplified crop rotations, but they also lacked the concentrated power of potent chemical fertilizers.[55] Commercial grain farmers had little use for the "organic" manure.

Many farmers struggled to rid their farms of this unusable waste. One approach was to simply ignore the problem. In the summer of 1977, inspectors discovered a shocking example of neglect—a dozen piglets found drowned in manure in their LPG near SZMK Eberswalde. The cause: overflowing storage containers. In other cases, farmers used the manure to literally hide their livestock losses. Inspectors discovered farms where dead animals were being thrown into manure lagoons.[56] These cases, while sensational, were rare. Many more farmers, as Pelta discovered, took care of the problem by dumping the toxic manure in the woods instead.

Many farmers, however, still clung to the notion that manure was valuable. In the 1979 section "Is the Soil Getting What It Needs," Pelta printed a letter from Herbert Lehmann, a farmer in the district of Herzberg. Lehmann described the increasingly common practice of manure dumping: "In many places, manure is thrown away clandestinely in the woods, very much like trash. Such practices are particularly infuriating, since we all know that the creation of more soil humus is one of our central tasks."[57] In another issue of the *DBZ*, Pelta published a picture of a field littered with untidy, unused manure, with the caption "What a mess! On a field in the LPG Grünlichtenberg, just off the F 169 highway between Karl-Marx-Stadt and Döbelin, a massive 'manure storage area' reaches probable record size, sprawling beyond the borders of this photograph."[58]

Pelta's picture and Lehmann's letter seemed to hold out hope that technology and determination would overcome these obstacles. Wrote Lehmann, "We livestock farmers treasure manure, despite the nature of this conundrum. . . . This issue of manure disposal is debated around here regularly, particularly amongst our cooperative partners in grain production. A sticking point for all of us is the issue of manure pits. The main problem for us is that in the past year, the size of our livestock populations has increased, while our manure storage capacity remains unchanged. As a result we have applied for a 240,000 mark reconstruction of this capacity."[59]

Lehmann was not wrong to hold out for state aid. The regime did invest in many of the country's LPGs. The three years following the implementation of the Grüneberg Plan saw the expansion of manure silo construction for LPGs across the GDR. Pelta took note of such a construction project near the city of Karl-Marx-Stadt in 1982. The LPG had already finished two new slurry pits, with plans for another solid manure preparation pad underway. Ideally, these storage facilities could handle the manure from both large and small facilities, eliminating the need for manure dumping in the woods and halting the destruction of fields.[60]

Construction, however, failed to keep pace with demand. A 1972 confidential internal report projected that by 1985, only half of the needed storage capacity would be met in the GDR. Instead, most farmers were storing their manure on unsecure land without adequate drainage ditches to catch liquid runoff. Farmers who attempted to stack manure on insufficiently developed yards or to remove it without proper equipment were only increasing the dangers of runoff and contamination.[61] By the early 1980s, the practice of *Zwischenlagern* had become the de facto standard for manure disposal. Reports of completely unregulated manure dumping were becoming routine. "The most important thing (for most farmers) is to keep their [manure] tanks empty," one person explained to the *DBZ*.[62] Some older farmers who had come of age when traditional stacking was still possible viewed the ZL as dangerous and wasteful. Heinz Krüger, a retired LPG member, wrote to the *DBZ* in March 1979 to say that "large piles of rocks and shoveled barn muck line the edges of many of our fields. Many of these areas serve as merely the dumping ground for *Zwischenlager*. Must it always be so?"[63] As the technological breakthrough failed to materialize, nitrates poured unabated into the East German environment. By 1980, the problem had become a full-blown health epidemic, and perhaps the most lasting legacy of East German agriculture.

POLLUTION'S TOLL

In late spring 1979, men in hard hats drove a mobile crane and large tractor trailer down to the shore of Plätlin Lake. The body of water was one of the "800 lakes" that dotted the famous Mecklenburg Lake District. Workers carefully maneuvered a twenty-two-meter-long cylindrical machine off the flatbed and into the water. Lying on its side, the object looked like a cross between a factory smokestack and a submarine, complete with periscope. Boats towed the machine out to the center of the lake and then flooded one end until the cylinder flipped into a vertical position, with the lower end close to lake bottom and the top end protruding above the surface. Once in position, the device rumbled to life. The machine was a massive aerator, designed to combat the algae blooms blocking out the sun and suffocating all aquatic life in the lake. East Germans had flocked every summer to the lakes for rest, relaxation, and relief from the summer heat. But over the previous ten years, the once famously clear waters had become opaque, slimy, and devoid of fish. Only the aerator could restore the beauty of the lake. Treating the lake as a giant aquarium tank, the lake-aerator system was designed to pump air to the bottom, oxygenating the water to help fish, amphibians, and helpful microorganisms restore the old environment.[64] Ironically, the technology had been designed for use in the lake-sized slurry lagoons of industrial livestock facilities like Eberswalde.

The connection between industrial livestock farming and polluted lakes was more than a coincidence—they were directly linked. Other bodies of water like Plätlin Lake were suffering throughout East Germany. Algae blooms and dead fish were symptoms of a change in water chemistry, known as eutrophication, a chemical reaction brought on by increased amounts of nitrogen and phosphorous in water. In a lake, for example, nutrient-hungry plankton and other plant-based organisms feed on these, which in turn causes them not only to reproduce at a greater rate, but also die in greater numbers. The algal population explosion deprived other water-based organisms of oxygen.[65]

Accelerated lake eutrophication was only one symptom of the factory-farm revolution. Mega-facilities like Eberswalde, as well as thousands of LPGs, released staggering quantities of nitrates and other chemical pollutants into the environment. Slurry lagoons held in suspension a time bomb of untreated and dangerous heavy metals such as copper, cadmium, and zinc.

Nitrate-laced surface water and aquifers were killing forests, wild animals, and even people. Elevated nitrate levels, a known risk factor for cancer and other health problems, were now spreading into the drinking water of East German cities. As the regime noted in a secret report, exposure to nitrate levels above the 40 mg/l level "increased the risk of damage to the brain, heart, blood vessels, liver, kidneys, and spleen."[66] Most traumatic of all was a rise in infant deaths, linked to blue baby syndrome, or methemoglobinemia, a condition in which nitrates inhibit the ability of hemoglobin to carry oxygen through the blood.[67]

By 1980, the Politburo was aware of the damage being wrought by nitrates after the MfUS circulated a detailed analysis of agricultural pollution. The authors offered a daringly candid assessment of the severity of the problem, underlining the catastrophic impact for the benefit of the regime's inner circle. While some ministries, notably the MLFN, lied about the causes of nitrate pollution, the ministers of the MfUS presented hard evidence in their report. They noted that between 1975 and 1980, the quantity of manure produced by industrial and traditional livestock facilities had increased from 48 million tons to 120 million. Simultaneously, the percentage of facilities operating without straw bedding had increased from 25 percent to 87 percent.[68] The highest rates of drinking water contamination (around 60 percent of the local population) occurred in areas that drew their water from surface sources, such as rivers, lakes, and reservoirs.[69] Nitrate contamination was not uniform. For example, in East Berlin, only 16,500 residents (or around 1.5 percent of the population) had contaminated drinking water, compared to the Dresden district where nearly half a million people or 25 percent of the population was affected.[70] The authors offered a final, damning conclusion. "Yes, this problem, as the literature shows, has become of increasing international concern. But in the GDR the danger of nitrate contamination, through the combination of our strained water supply, and dense development of grain and animal production, has become more critical than in any other European country."[71]

While some East Germans discovered evidence of manure pollution on hikes or while camping, others first noticed physical evidence in their sinks and showers. In a letter to the minister of environmental protection, Winfried L. detailed a yearlong ordeal struggling with the contamination of his drinking water by a large neighboring LPG T. The collective farm near Gera opened in October 1979 with twenty thousand cattle.[72] By the following spring, fecal matter from manure began to show up not only in the little brook that ran behind his house, but also in his own well. As the farm brought more of its liquid manure out for field irrigation, several other private wells became contaminated, according to Winfried L. The mayor of the village called the Department of Health Inspection, which immediately closed the twelve wells. Local authorities implemented tests, ordered the wells cleaned with chlorine, and tested them again. Despite all these efforts, the health inspector detected more fecal matter in the wells at the end of July and ordered that the wells remain closed. Living with his partially disabled wife, Winfried L. worried that he wouldn't be able to keep up with household chores without direct access to water. He implored the minister to use his influence to reduce manure contamination and secure fresh water for the town.[73]

Despite mounting evidence, the MLFN dragged its heels. Appended to Winfried L.'s plea was a note from Minister of Agriculture Kuhrig to Willi Stoph, chairman of the Council of Ministers. It illustrated the regime's complete lack of urgency. While Kuhrig acknowledged the growing threat of manure contamination as detailed in Winfried L.'s letter, Kuhrig saw no need for further action until *all* sources of water pollution had been identified. Only then could a truly comprehensive plan of action could be developed, Kuhrig wrote.[74]

Drinking water contamination was not just deadly but also expensive. As the problem spread through water tables and fresh water sources, more wells like Winfried L.'s had to be closed and a substitute water supply provided at great cost to the government. In 1979, the drilling of new wells and installation of new pipes cost DM 3.3 million just in the district of New Brandenburg. In the same year, the government reported providing clean water for infants in 263 villages in the district of Gera, and some five thousand more infants in the district of Dresden, at an undetermined cost. In the area around Cottbus, the regime was supplying 170,000 bottles of water a year for families with additional deliveries for babies in an additional twenty-six villages. In 1979, the MfUW estimated that the special water deliveries were costing the regime 250 times the annual budget for the entire drinking water system of East Germany.[75]

The pollution crisis also affected plants and animals. In the area around Eberswalde, nitrate pollution caused forest dieback for miles in every direction. Eight hundred hectares of pines and spruces had been destroyed by 1990, the product of an annual release of 3,500 tons of nitrogen into the atmosphere.[76] At the nearly identical SZMK Neutstadt/Orla, the VEB Forestry Project out of Potsdam uncovered a slow-moving *Waldsterben* (forest death) disaster. In the catchment basin surrounding the complex, the affected

FIG 4.2 Brimming manure tanks at SZMK Neustadt/Orla, February 16, 1990. Photograph courtesy of BArch, Bild 183-1990-0216-031/Jan Peter Kaspar.

forest area grew from 219 hectares in 1982 to more than 806 hectares five years later.[77] The two hundred thousand hogs there also damaged the air quality of a city with six hundred thousand residents.[78] In 1990, *Der Spiegel* visited the forests near Neustadt to survey the devastation. "Ammonium nitrate has killed pines and spruces, while new plantings died almost immediately after. Where once a timber forest stood, now only a wasteland exists—no trees, no shrubs, only manure to the horizon."[79]

Environmental experts worried about effects of the factory farm on local ecology. Nitrates not only killed off native species, but also encouraged hardy pioneers, or ruderals, to take their place. Furthermore, the development of ever-larger fields and pastures reduced habitat for other wild species.[80] Industrialization, ministry officials argued, risked precipitating a dangerous cascade of disturbances throughout the environment. "Already a new succession is underway in salt marshes, prairie grasses, heaths and moors, each home to a diverse array of animals, particularly birds, like shore-species curlews, black-tailed godwits, sandpipers, redshanks, and terrestrial birds like black grouses, woodlarks, and pipits." The ministry report continued, "Taken together, it is clear that the general eutrophication of aquatic and terrestrial ecosystems, and the increased usage of pesticides in agriculture and forestry have led to a considerable decline in species diversity. . . . Numerous studies exist now that show wherever the stability of an impoverished ecosystems has weakened greatly, these areas are increasingly vulnerable to external and internal disturbances, their mechanisms for self-regulation no longer intact."[81] Even in notoriously dogmatic East Germany, ecological ideas and concepts became useful ways to critique the adverse effects of rural development.

CONCLUSION

East Germany's manure crisis offers an opportunity to rethink the relationship between state socialism, global capitalism, and agricultural pollution in the late twentieth century. Over the last thirty years, the East German story has become a familiar one throughout the developed agricultural world. Critics who once decried the destructive gigantism represented by Eberswalde overlooked the pervasive and ongoing epidemic of nitrate pollution from industrial livestock-keeping today. In early twenty-first-century America, manure runoff continues to affect communities from Tulare County, California, to Toledo, Ohio.[82] Consider rural China, where industrial hog farming has turned the country's fourth largest lake into an open sewage lagoon, killing fish and burning the skin of anyone who dares touch the surface.[83] East Germany's situation was hardly unique. So what was it that tied the East German experience to this global history of agricultural pollution? It was capitalism.

As we have seen, the GDR based its agricultural development on an American model of industrial farming. By the late 1960s, without access to the world of cheap inputs such as grain, labor, and capital, the East German factory farm faltered.[84] It was "rescued" by changes in the 1970s to global capitalism and Erich Honecker's turn to the West. While the first secretary believed cheap credit and grain would accelerate the transformation of the country into an export land, the shift pulled the country's pork and pigs into global flows of capital and commodities. What few seemed to realize, however, was that the East German reliance on Western markets brought a faster and more terrible environmental reckoning than there would have been without the turn toward the West.

Debt and manure pollution were not separate crises. They were two sides of the same coin—the high cost of cheap food. The sheer scale and concentration of East German hog industry, which reached their zenith in the early

1980s, reflected the regime's growing dependency on capitalist sources of fodder and credit. As the cost of those inputs grew, the regime raced to break their dependency through technical fixes, like the creation of Eberswalde, the implementation of the Grüneberg Plan, and the manipulation of pig digestion and social behavior. Each "fix" doubled-down on the underlying premises of industrial agriculture—planners "needed" bigger factories, or higher concentrations of pigs, or cheaper and greater volumes of grain—a strategy that only made toxic pollution worse.

Of course, the regime went into industrial pig production with its eyes wide open. The MfUS's warnings were not enough to get them to change course. From the perspective of the planned economy, this was understandable. Reform would require a dramatic intervention—tantamount to the repudiation of thirty years of industrial development. Perhaps this was an impossible choice. As the MfUW would argue, it was a question of economic production versus public health. On the one hand, agriculture was "inconceivable without nitrate fertilizer." On the other hand, "drinking water represents an essential provision for the people, requiring at a minimum certain basic qualities."[85] In this way, the East German state, much like capitalist states then and now, confronted an impossible choice: cut back on production—and thus curtail the pollution epidemic—or press forward with the status quo of cheap food.

As the 1970s drew to a close, the contradiction between industrial agricultural production and deteriorating public health grew starker. While the regime chose to ignore the problem, many East Germans could not. Throughout the decade, citizens like Fridger Pelta bore witness to the growing problem of concentrated livestock production. In the early 1980s, some members of the regime decided that the time had come to end East Germany's dependence on the West. The GDR would have to cut its foreign grain imports, and in doing so cut the reliance of its agricultural sector on western credit. All of this came to head in 1982, the year of Pelta's mossy giants. Instead of freeing the country from this system, the country's industrial pig population crashed, setting off a cascade of shortages that reverberated through nearly every sector of the economy.

The ensuing pork shortage of 1982 would ultimately threaten the GDR with insolvency as a nutritional and epizootic crisis for animals turned into a political crisis for state socialism. Toxic animal waste played a direct role in precipitating the crisis, and it became the symbol of the fundamental problem of overproduction and reliance on Western markets. As we will see, the pork crisis of 1982 tied together not only the problems of manure disposal and fodder production, but, in the end, it forced Western creditors to bail out the East German regime. For ordinary East Germans, however, it revealed the inadequacy of their government. It also revealed the importance of a new kind of private, subsistence farming to fill in the gaps.

CHAPTER FIVE

PIGS IN THE SMALL GARDEN PARADISE

ERICH HONECKER HAD A RELATIVELY SHORT COMMUTE TO WORK. EVERY day, the East German first secretary climbed into his Volvo limousine in the secluded wooded settlement of Wandlitz and was driven by his chauffer twenty miles down the highway to his office on the banks of the Spree River opposite the recently constructed Palast der Republik. In early 1977, outside the town of Bernau, Honecker noticed during his commute an untidy collection of cinder blocks, tin roofs, and metal fencing lining the side of the highway. The encampment seemed to have erupted overnight in the middle of a fallow field, not far from his residence. The unruly settlement must have displeased the East German leader, and so he asked the ministers of agriculture and construction to check in on the new garden colony, first in 1977, to see if the materials obviously "liberated" for construction had undermined the capacity of other industries. A report from that year stated that everything was still above board, momentarily assuaging his concerns. The following spring, as the motley assemblage sprouted more wings, Honecker became frustrated. On May 8, 1978, he ordered his Wandlitz neighbor and chief economic official, Gunter Mittag, to find out who was responsible for the monstrosity and whether or not the settlement was wasting agricultural space.[1]

Unfortunately for Honecker, the problem was mostly one of his own making. According to the report, three agricultural officials had authorized the construction of a garden colony near Wandlitz in April of 1976.[2] In addition, the author of the report attached an excerpt of a resolution entitled "Measures and Assignments for the Promotion of the Work of the VKSK (Union of Small Gardeners, Settlers, and Small Animal Breeders) of the GDR and the Initiatives of its Members," passed on August 3, 1977, by the Secretariat of the Central Committee—a body overseen directly by none other than Erich Honecker.[3]

Honecker could be forgiven his confusion—after all, he seemed himself unsure where the individual production of gardeners and collective farmers fit in the future of state socialism. Just two years prior to Honecker's run-in with the Bernauer garden colony, economic planners seemed on the verge of eliminating individual production entirely. In December of 1975, as planners prepared farmers for the impending implementation of the Grüneberg Plan, the state also announced its intention to lower the 1976 purchase prices for all individually produced animal foodstuffs, like rabbit meat, eggs, and poultry. At the new price levels, many animal keepers would simply no longer be able to afford to keep such animals on their own. The move would hit retirees especially hard, as many used such practices to supplement their meager incomes. Many collective farmers, themselves small individual producers, were apoplectic about the reform, with some wondering aloud, "Do we now produce so much meat that there is no longer a need for individual animal keepers?"[4] For Honecker and the Politburo, the answer would certainly have been yes.

A decade later, Honecker's confusion had dissipated. An East German PhD student named Klaus Siegemund cited the general secretary, and his speech at the Eleventh Party Congress of 1986, as an example of the prominent and historical position of garden culture in the GDR. "Also in the future, the individual production of the personal household economies of our collective farmers and workers, as well the [VKSK] will receive our full support. They have secured a place in the balance sheet of our economy and effectively supplement the production of society over the long run."[5] So what had changed?

Honecker's 1986 support for garden farming had been neither the paragon of consistency that Siegemund believed nor the product of a massive conversion or noetic moment. Rather, it emerged begrudgingly in the years following the planning and implementation of the Grüneberg Plan. For much of East Germany's history, the regime had been ambivalent, if not hostile, to gardening, believing it was either a remnant of the "bourgeois" culture or a reminder of the food shortages that had plagued Germans after both World Wars. The year 1976 was different. Once the Grüneberg Plan was enacted, the regime abandoned the view of gardening as a threat to socialist production and at the Ninth Party Congress formally endorsed the practice as a worthy socialist recreational activity.[6]

The decision freed planners and gardeners alike to strengthen the activity, which became enormously popular over the next decade. The amount of land cultivated by individual households in that period nearly tripled, from 240,000 hectares to 630,000 hectares.[7] Members of the VKSK, the mass political organization that oversaw the more than 830,000 garden plots in the country, grew from 800,000 members to 1.49 million, and worked on an estimated sixty thousand hectares of land.[8] Another 3.5 million GDR citizens engaged in some way with the activities of the VKSK on a daily basis by 1985, in addition to seventeen thousand East Germans who claimed membership in a self-sufficient *Sparten* settlement.[9] Those lucky enough to acquire a garden plot treated them as their own private refuge. There they would relax with family and friends, garden in good weather, and enjoy the fruits of their own labor—a general feeling many represented with a short aphorism, carved on a piece of decorative wood—*Klein aber mein*, or "small but mine."[10] Taken together, this diverse group of individual producers, nearly one in three East Germans, cultivated more than 10 percent of the country's arable land. In the words of historian Isolde Dietrich, the GDR was a veritable "small-garden paradise."[11]

German gardening, however, included much more than fruits and vegetables; it also involved animals. At the apex of this system was the garden pig, or a pig raised by East Germans on small, individually managed plots of land. Individual producers could sign contracts to raise pigs and sell them back to a state slaughterhouse for additional income or were permitted to slaughter them privately for family consumption, outside the planned economy. While biologically indistinguishable, the garden pig diverged most from the industrial pig when it came to its environment. It was reared in the company of only a few pigs, in addition to other small and large livestock. It never experienced the crowding of specialized facilities. Its diet was varied, not just grain rations of wheat, rye, and barley, but also farm family leftovers, spoiled produce, and stale bread. While these were differences largely of treatment, they

FIG 5.1 Garden colony in Eisenhüttenstadt, 1982. Photograph courtesy of BPK/Interflug-Luftbildarchiv/Art Resource, NY.

were not superficial. They contributed not only to the superior health of the garden pig, but also its massive size.

In this chapter, the garden pig stands at the top of a broad and varied system of small-scale agrarian activities and forms of land tenure. Garden farmers not only grew fruits and vegetables on their plots, but also raised a variety of animals for their fur, eggs, and meat. Small livestock such as chickens, geese, rabbits, and nutria were as common in East German backyards and city gardens as they were on farms. Many kept bees. On collective farms, farmers had more space and worked a small allotment, generally a half hectare, for their own private use. They raised their own meat (often a couple of sows with piglets, goats, sheep, rabbits, and even a few cows), grew their own fruits and vegetables, and basically followed the traditional tenets of *Haushaltswirtschaft*, or rural home economics. While not allowed in urban gardens, the garden pig was the perfect symbol of East German garden culture. It remained, on the one hand, rooted in the traditional, pre-industrial agricultural past, while at the same time linking the small-scale free-time activities of millions of East Germans with the industrial system of the present.

The East German garden paradise was never planned. While the regime had remained ambivalent since 1949, everyday East Germans pushed for its expansion. Paradoxically, it wasn't until the East German turn toward the West that small-scale agrarian subsistence became inscribed within the planned economy.[12] As food production increasingly prioritized the demands of Western markets, breakdowns in domestic supply became more frequent. Periodic shortages in 1976, 1978–79, and most disastrously, in 1982, reaffirmed the importance of subsistence gardening. In response, planners set aside more land for gardens and increased financial support and subsidies for individual producers. New *Aufkauf* (purchasing) stations in retail stores, grocery stores, and weekly vegetable markets formed a produce collection and distribution network for small producers. It all culminated in the 1984 agricultural price reform, a law that granted a formal right to private production. While Erich Honecker may have once scorned the practice, everyday East Germans convinced him and his cohort of two things: first, that garden farming belonged under state socialism, and second, that factory farming could not function efficiently without the flowers, fruits, vegetables, bees, rabbit, and pigs of hundreds of thousands of garden farmers.

GARDEN CULTURE AND THE INDUSTRIAL REVOLUTION

Since he was a child, Wilhelm Jakob had known of the restorative power of labor. A life without it seemed nearly inconceivable to this son of a dairyman. Labor had structured his childhood in the small village in Brandenburg where he grew up. It had helped his family prosper years later on the northern outskirts of Berlin. Even at the age of eighty, Jakob still worked with his hands. Every day, he returned to his small garden plot in the Schüßler Garden Colony of Pankow, to weed the vegetable beds, dig up the potatoes, and carefully prune his cherry tree.[13]

This garden, however, was much older than even Jakob. Its name, Schüßler, had come from the land's original owner, who around 1900 established a tavern on the northern edge of the city. Berliners flocked to the tavern on warm nights to drink, dance, and take a break from the summer heat. The First World War dissolved that world. The garden on both sides of the Panke River came into being during the winter of 1918, feeding families through the upheaval of civil war and hyperinflation and another economic crisis in the 1930s. Even as fascism "burned the world anew," the garden served as a refuge to families seeking shelter from nightly air raids.[14]

The story of Jakob and the Schüßler Garden Colony appeared in the East German weekly magazine *Die Wochenpost* in late 1978. The author, Sigurd Darac, used the long life of Herr Jakob to connect the tradition of German garden culture in East Germany to the not-too-distant past of 1900. From Darac's perspective, East Germans were heirs to a specifically German garden legacy—a legacy that had staved off famine and reconnected generations of families to each other and to nature. This was not entirely false. Gardens played a central role in the rise and expansion of European cities. As governments grappled with the effects of the industrial revolution—particularly the displacement of rural populations and the growing problem of pauperism—many principalities turned to gardens for their capacity to feed.[15] For Germans, the major period of pauper garden establishment was in the 1830s and 1840s, when Kiel in the kingdom of Schleswig-Holstein (which at the time belonged to Denmark) established its first pauper garden (or *Armengärten*), followed by Königsberg, Frankfurt am Main, Dresden, Berlin, and, most importantly, Leipzig.[16] It was in this last city that the German allotment garden movement arose and spread under the name Schreber Gardens among bourgeois denizens of Leipzig as a social critique of urban environments.[17] Garden culture also fit into a modernist view of urban design. While East Germans leaders were never conscious of his work, the turn-of-the-century German landscape architect Leberecht Migge presciently anticipated the type of country East Germany would become.

The main thrust of Migge's work revolved around the role of gardens in the form and function of modern cities. As opposed to his famous forerunner, Frederick Law Olmsted, whose designs treated parks as separate arcadian spheres through which to escape urban life, Migge believed that landscape design placed people back in their landscapes. His most influential and important work was written under famine conditions toward the last year of the First World War in Germany.[18] The infamous "turnip winter" of 1917–18, when the eastern and western fronts cut Germany off from foreign trade, grain, and the raw materials of food production (like fertilizers), forced many Germans, especially in Berlin and Hamburg, back to the Schreber gardens of the nineteenth century.[19] The trauma of food scarcity convinced Migge of the importance of gardens and their ability to produce food in modernist urban design.

In two works published after the war, Migge laid out a synthetic vision of a German "city-land" settlement, where everyone would have the land necessary to produce their own food. The first of these works, *Jedermann Selbstversorger!* (Every man self-sufficient), focused exclusively on the design and function of garden plots. Influenced by the works of Petr Kropotkin, Migge stated, "My intention is to show how a family with a modest garden can pay for the land with their own handwork, and generally support themselves as well."[20] In the same year, Migge also published a philosophical treatise on his principles of modernist design, known as the Green Manifesto. In it, Migge synthesized more than two decades of design work into a formal overhaul of the Garden City movement and other attempts to reform the city. The manifesto connected many of the prescriptions of *Jedermann Selbstversorger* into a reorganization of Germany's national space. This new Germany would no longer be dominated by a rural/urban division, but instead "bring the city back to the land" and "create city-land!"[21] City-land would free the working class from the crushing pressure of centralized capitalism through the single concept of a self-sufficient economy. Self-sufficiency in this model originated in the individual, was amplified through cooperative communities, and then again once more through a larger city. This inverted pyramid of agricultural production elevated the labor of gardening from a crude form of social welfare into the restorative foundation of a new, postwar Germany.[22] Migge's manifesto also outlined the German Democratic Republic that was to come: a country filled with gardens where citizens raised not only their own fruits and vegetables but also their own livestock and animal products, and where a careful balance between rural and urban space left room for "garden belts" and self-sufficient "settlements."

EAST GERMANY, "CITY-LAND"

Although East Germany's planners never heeded Migge's call for the creation of city-land, the postwar history of rural and urban development in the GDR created the material conditions for a country that reflected Migge's ideal. Much of this had to do with the state of the country at zero hour, 1945. The war had devastated agriculture in the Soviet Zone of Occupation. In May 1945, grain production only reached 15 percent of its 1939 level, and by December, it had risen only to 25 percent. Through 1947, the food supply barely improved, as a severe winter in 1947–48 ended the relative progress of the previous eighteen months. Throughout this period, gardens provided bare subsistence to Germans living in the Soviet Zone. In Berlin, where the population in 1945 hovered around 1 million, some 250,000 gardens were planted wherever space could be found. As historian Donna Harsch has argued, this foundational moment in the GDR's history embedded food scarcity as a latent neurosis in the national consciousness.[23]

In those early years, self-sufficiency was a priority for East German families, much as it had been for Migge in the wake of the First World War. Allotment gardens and the small plots created by the Soviet land reform of 1948 provided many families with the land to feed their families. The GDR regime kept track of such activities from the very beginning, designating them under the category of "individual production" in planning records. "Garden" production generally encapsulated the cultivation of only a few hundred square feet of land, where people grew primarily vegetables and fruit trees but also kept smaller animals like chickens, geese, and rabbits. Household production (*Haushaltswirtschaft*), on the other hand, resembled more closely the mixed production of traditional farming. It included the cultivation of garden produce and more land-intensive fodder crops (like root vegetables, grasses, and grain), small herbs, and larger livestock such as cows and pigs.

Individual production firmly rooted East Germans to the places they lived. While changes in development, economic, and education policy attempted to induce people to move to the cities, by the 1980s, the GDR had largely given up on the goal of a mostly urban society.[24] The focus on the development of a strong, well-educated working class in the countryside, especially after the end of collectivization had created rather stable conditions. As a result, the demographic distribution of the population was in many ways perfectly balanced between the city and the country. The percentage that lived in cities larger than one hundred thousand residents barely rose between 1950 and 1985, from just over 20 percent to 27 percent. The largest city, East Berlin had a population of over a million, while the two other largest cities, Leipzig and Dresden, had a half a million each. They were the exceptions. Most cities were modest in size. In contrast, a remarkably large percentage of the population—over twenty-three percent—lived in hundreds of villages of two thousand people or fewer.[25] Even the residents of East Germany's cities still had access to land in one of thousands of garden colonies, of which the highest numbers were located in Karl-Marx-Stadt (1,610), Dresden (1,345), Halle (1,226), and in Leipzig (992).[26] The demographic distribution and agrarian forms of the country seemed to follow some unspoken mandate of Migge's to create more city-land.

DUAL INCOMES AND *HAUSHALTSWIRTSCHAFT*

Individual production, or *Haushaltswirtschaft*, in the countryside took a different form. Its primary objective had been the maintenance of the rural family, and it had existed alongside a variety of European forms of land tenure, from the estate farming of Junker nobles to the commercial agriculture of large individual landholders. It supplemented rural family incomes by providing food with a small surplus for sale at the market. Individual production continued in the GDR. Collectivization, while originally designed to dissolve all capitalist forms in agriculture, preserved individual production. It helped to stabilize rural households during this period of intense economic, social, and political upheaval.

Each collective farm (LPG) charter guaranteed a general-use right of land for all members, granting each farmer a half hectare of land (about one and a quarter acres) for private production. These charters permitted each member to raise up to two cows, two sows and their offspring, five sheep, and as many other small animals (chickens, geese, rabbits, goats) as they desired.[27] The granting of use rights during the first collectivization drives were not merely a sop to those farmers still clinging to their desire for private property rights. It was intended to provide a level of subsistence below which rural farmers were not allowed to fall—a crucial guarantee from the state after the lean years of the postwar period.

By the 1970s, nearly a decade after the completion of collectivization, planners continued to promote individual production. This was a means to fulfill the demands of often under-serviced rural communities, since industrial development had oriented the majority of fruit and vegetable production toward urban centers. Cities were always supplied before the countryside. This hierarchy of delivery needs left many rural communities with only slim pickings in the produce aisles through much of the year. *Haushaltswirtschaft* provided fresh produce when long delivery routes from the fruit and vegetable cooperatives could not.

Individual production also supplemented the incomes of marginal community members, such as retirees and women. Communal farms and regional councils routinely granted these members of the community direct access to plots of land, while also helping them form production contracts with other "socialist" buyers, like cafeterias and schools. The benefits spread both to the gardeners, who gained a second income, and to rural cooks and consumers, who enjoyed better ingredients in their meals. Although the state compensated women and retirees for their labor, they officially designated individual practices as a free-time or recreational activity.[28]

The secondary status ascribed to individual production belied its greater significance. Most farmers found ways around the state's categorization of such work as a free-time activity. For example, the law required farmers to put in eight-hour days in their LPG. Many managers and chairmen recognized the necessity of such "hobbies" and allowed more flexible work hours for their members. Because of these legal requirements, a large portion of individual household production was divided by gender. Since many women worked from home during seasons of low activity, a large portion of this labor fell on wives and grandmothers who "found" the time to check up on the chickens, weed the garden, and feed the family pigs in moments of inactivity.[29] There were other ways of taking advantage of the collective farm by-laws in order to support individual household production. For example, every LPG member, from the newest members to production managers, was required to weed a quarter of a hectare of beets by hand once a month, since the tractors always left some weeds behind. In exchange for this work, the LPG paid each farmer approximately one ton of grain. An enterprising individual producer would often do the work of his fellow farmers in exchange for their share of the grain, which thus allowed him or her to raise more animals. Individual producers were also known to cobble together the half-hectare plots of many neighbors into a single large plot, traded for other goods or a percentage of their second income.[30]

Individual production also had deep cultural ties. Many collective farmers saw the practice as their own personal legacy while also enjoying the perks of individual production. Rolf Urland, an LPG member from the district of Sondershausen near Erfurt, described how his father, during the lean years of the postwar period, had fed his family in this manner. Although by the 1970s, subsistence needs no longer shaped rural life, the practice of household agrarian production provided a crucial connection for Urland and his family to the work and traditions of his father. More than any anything, Urland loved raising his colorfully striped Amrock chickens in his backyard shed.[31] In language familiar to anyone who has eaten food they have grown themselves, he described the joy he got from eating their eggs: "One can't disdain a fresh egg on his own table and a tasty roast chicken with it."[32] Urland reveled not only in the direct connection between his labor and his food, but he also earned a significant amount of money from his "hobbies."

> A pig of a good-sized weight class brings no less than a thousand marks when we deliver it. For a few other agricultural products, the prices were also recently raised. . . . Our family sold in this year alone 3000 eggs, over 20 kilograms of wool, and four hogs with a weight of 170kg each. Taken together we collected a pretty pile of cash—all earned honestly. The only cost was energy, effort, and sweat . . . our family covered our own demand for meat through household production, and with the remaining four hogs we raised and sold, fed an additional eight people from the city.[33]

Urland managed to fatten all of his animals through a mixture of individually raised fodder he purchased from other small producers, and cheaply subsidized fodder from the state. All the extra, secondary work, however, did not prevent Urland from going on vacation with his wife. Since the work required was minimal, he could direct family members to keep an eye on things while he and his wife visited the People's Republic of Hungary, as a reward from the LPG.[34]

GARDENING DURING THE ULBRICHT ERA

Like *Haushaltswirtschaft*, garden culture in the first two decades of the GDR remained a marginal activity, largely due to the efforts of the regime. Even though gardens had helped sustain the postwar populations in the Soviet Zone, both the occupation authority and the first leaders of the GDR viewed gardens with skepticism. Walter Ulbricht worried that they could be used as a fifth column to undermine the authority of the newly enshrined German communists.[35] Others, like planner Lothar Bolz, believed gardens were tools of capitalist exploitation, turning "the working man into a rabbit breeder or cauliflower planter, but under no circumstances a participant in political demonstrations."[36] Depending on the perspective, gardens had the power to rile up the proletariat with false ideas or lull it into complacency. In either case, it was best if gardens had as little power as possible.

While the Socialist Unity Party (SED) did not ban gardens, they marginalized them. Initially, they placed garden regulation under the bureaucratic jurisdiction of the Free German Trade Union (FDGB), East Germany's largest mass organization. The FDGB, however, had little time for gardens, as they were only a small part of their broader portfolio of social services.[37] The regime even flirted with plans to collectivize the "colonies" in the 1950s but retreated from the position by the end of the decade as austerity ended and gardeners pushed for formal recognition by the state. In 1958, the Politburo relented, and formally ordered the Ministry of Agriculture along with regional functionaries from various agricultural sectors to explore the creation of mass party organization for East Germany's gardeners. The organization, the VKSK, was formally established on November 28–29, 1959, and initially had more than 850,000 members (membership was mandatory). It gave garden culture nominal recognition among the social politics of the GDR.[38]

The VKSK was the large umbrella organization that oversaw all forms of private production (except *Haushaltswirtschaft*). Similar to all "parties of mass organization," it was divided into national, regional, district, and village offices. To complicate matters, many of the geographic distinctions gave way to technical specializations as well, such as VKSK units for bee keeping or rabbit breeding—the "small animal keepers" of the group. The "settlers" inside the VKSK were also called *Sparten*, or self-sufficient settlements. They resembled garden colonies, but generally organized the local production of fruits, vegetables, and animal products in small villages.[39] Taken together, the VKSK's mission was to incorporate the efforts and social organization of its members into the broader, albeit vague project of "the construction of socialism."

Despite the formal recognition of the party, the 1960s did not see the regime fully embrace garden culture. The VKSK still remained a near nonentity in the Council of Ministers and Politburo, since from the planners' perspective, gardens added little economic value to the plan. Many in the regime still suspected gardens as a refuge for bourgeois culture and capitalist sympathies. Such views were prevalent. As one advice seeker wrote to the woman's magazine *FÜR DICH* in 1975, "I have a garden. Does that make me snobbish?"[40] In the letter, the magazine subscriber, Marlis Allendorf, related a disagreement her friend Regine had had with a colleague at work. Regine had been telling her coworkers how excited she was to head out with her family to her *Grundstück*, or little plot of land, for the weekend. Her colleague, who was her office's party secretary, pointedly commented that it must be nice to have a "summer house"—snidely labeling her garden as one of the trappings of bourgeois culture. Regine could only respond that little about her time in the garden was bourgeois; in fact, her weekend trips were filled with hard work. "It is definitely not the 500 square meters that draws me there—but rather the beauty of seeing the coming and going of spring, summer, and fall outdoors; the workout gardening gives my body; and the fact that I don't need to constantly follow the kids and keep them in line. And when we arrive home on Sunday evenings, I have the hands of a hard laborer and can still feel the physical exertion in my back."[41] In other words, the only thing that redeemed the bourgeois garden was hard work, or as was the saying of the day, "A beautiful garden is a productive garden."

Nevertheless, ambivalence toward gardens persisted. During the 1960s, the regime pushed to dissolve many, alarming both leaders of the VKSK and city residents. More than ten thousand garden plots (more than 437 hectares) were rezoned for housing construction in Berlin alone between 1960 and 1972. The removal of gardens went hand in hand with the elimination of farming from the capital city. As late as 1975–76, the regime was still tearing down barns in the neighborhood of Lichtenberg to make way for new apartments.[42] Urban renewal displaced the allotments of some nine thousand families.[43] Even in the first years of the Honecker era, many believed the garden culture was on its way out.

Things changed with the Ninth Party Congress and the regime's focus on raising standards of living. Taking their cues from Honecker's promise to improve social programs, in early 1975, Berliners petitioned their government for greater access to green space. Not only did citizens demand an end to the dissolution of gardens, but they also requested that authorities establish regulations to protect urban garden space from development. In the formulation of urban residents, gardens were an essential right under state socialism, on par with Honecker's other social policy initiatives like cheap rent, food, and consumer goods. East Berliners requested greater access to "weekend plots" in Berlin suburbs like Köpenick, where they would have more opportunities for recreation and relaxation.[44] The following April, in reaction to the growing number of petitions, Gerhard Grüneberg designated a thirty-hectare field north of Berlin near Bernau for the construction of new garden colonies—the same garden colony "eyesore" that would bother Honecker during his daily commute.[45]

The construction plan proved perfectly timed, as the summer of 1976 was especially hot. Droughts followed, endangering the fall harvest, and many Berliners sought refuge from the heat in the existing gardens of friends and family. Honecker, recognizing the changing political climate for gardens, formally announced an end to the "liquidation" of East Germany's gardens in a September speech before SED party members. The party newspaper, *Neues Deutschland*, quoted Honecker as saying, "The development of the city must take into better consideration how the construction of every residential area fits into its surroundings. . . . We must quit the expropriation of gardens from small gardeners and stop cutting down trees with impunity. The environment should be protected and constantly improved."[46] Honecker and the rest of the Politburo could no longer ignore the growing demand in Berlin especially for garden plots and private space. Although he would forget, Honecker had brought garden culture back from the dead. He had also unwittingly left the door open for an expansion of the productive capabilities of gardens that would become a critical support for the industrial planned economy within the decade.

RECREATIONAL GARDENS AND "VEGETABLE BELTS"

The expansion of garden culture in the mid-1970s matched Erich Honecker's governing priorities—especially his Unity of Economic and Social Policy.[47] The first secretary desperately wanted to consolidate his power by putting as much distance as possible between his rule and the austerity of the republic's first two decades. He did this by focusing on raising "consumption" and encouraging citizens to seek out recreation, hobbies, and free time. Hoping to improve his standing in the eyes of the people, Honecker overlooked his reservations and endorsed the practice of gardening. The rest of the government got in line quickly. In 1977, the powerful State Planning Commission (SPK) ordered the Council of Ministers and the VKSK to establish formal "regulations for the state support of the development of small garden complexes," which when published in April, explicitly claimed that gardens promoted the "relaxation and organization of one's free time." The guidelines defined the form, duties, and rights of small gardens, garden colonies, and small garden parks, while only obliquely mentioning their productive capabilities, referring to "soil type" and its relation to the "intensive use" of the garden. Such language left the issue of usage up to the discretion of the gardener. Most significantly, the new regulations allowed gardeners to sleep "overnight" in their garden houses on weekends, thus creating a valued "second home" for East Germans used to the confines of enormous apartment complexes.[48]

While gardens would improve people's social lives, Honecker believed cheap food would improve their economic outlook. It was this faith that opened a loophole for the return of garden farming. Nowhere was this more apparent than in Berlin. At the exact moment that Honecker and the regime formally sanctioned the use and expansion of urban gardens, they also made plans for the full industrial provisioning of Berlin with fruits and vegetables. In 1974–75, planners announced the creation of a new "vegetable belt" to supply the city. This "belt" combined a group of LPG, state-owned farms (VEG), and collectivized garden cooperatives (GPG) into a productive unit and resembled in many ways the "green belt" of Migge's "land-city" settlement ideal of 1920s. The creation of the vegetable belt must have also put Honecker's mind at ease about the expanded access to garden colonies. With the Grüneberg Plan, the ramped-up production of Eberswalde's pork facility, and the vegetable belt, the future of food production appeared to be in an industrialized countryside, not in urban gardens.

Planners believed the vegetable belt would solve the two main problems for fruit and vegetable production: distance to market and perishability. Since fruits and vegetables begin to spoil almost immediately after harvest, the sooner they are sold to consumers, the better. And the only ways to slow this process are either through refrigeration or to deliver the food to the consumer as quickly as possible. The latter can be accomplished through an acceleration of the delivery speed or a decrease in the total distance to traveled. Since refrigeration technology was notoriously unreliable and inefficient in the Eastern Bloc, East German planners attempted to deliver produce faster.[49] Instead, the vegetable belt reduced the distance traveled by organizing direct deliveries, streamlining transportation, and locating production at the outskirts of the city.

Planners also had to contend with the surprisingly diverse and evolving tastes of Berlin's consumers. For several years, citizen petitions (or *Eingaben*) had clamored for more fresh fruits and vegetables in the capital city. By the mid-1970s, however, Berliners also demanded changes to the structure of their shopping experiences. They wanted to have the option of self-service in the grocery store, selecting their own produce by hand rather than being served by a salesperson, and they desired more "kitchen-ready" or prepackaged foods that they could prepare at home, as East Germans increasingly ate their meals in private.[50] Berlin was also becoming an increasingly sought-after destination for tourists from both Eastern and Western Europe, all of whom held their own culinary expectations. Planners like Bruno Kiesler recognized these evolving patterns of consumption and repeatedly stressed the need for greater supplies and wider variety of green produce like carrots, celery, parsley, soup greens, salad, winter vegetables, and other root vegetables, not to mention more exotic types of produce, like melons, red peppers, and garlic, which were not grown in the GDR in 1975.[51]

In 1974, the twenty-fifth anniversary year of the GDR, planners brought the vegetable belt into operation. The belt, however, was not merely a spatial designation for specialized vegetable farmers around Berlin, but also mechanism that tied together their production through a single, socialist middleman. The constitutive members of Berlin's vegetable belt—the chairmen of collective and state farms, collective gardens, and garden nurseries—incorporated this body, naming it the ZGE (*Zwischengenossenschaftliche Betrieb*) Berlin Fruit and Vegetables. The ZGE took over the difficult work of collecting, processing, and distributing the city's produce from every direction and integrated that produce into the large network of grocery stores and cafeterias. From the eastern towns of Wollup, Golzow, Gorgast, and Alt-Tucheband, the ZGE gathered greenhouse vegetables and red and white cabbage. From the central and north central neighborhoods of Bucholz, Elisenau, Werneuchen, Marzahn, Pankow, Weissensee, and Wartenberg came parsley, kohlrabi, savoy cabbage, white cabbage, kale, carrots, and lettuce. On the other side of West Berlin, large scale LPGs and GPGs organized production in the western villages of Nauen and Markee for kale, beans, leaks, brussels sprouts, celery, and cucumbers.[52]

The ZGE Berlin Fruit and Vegetables simplified not only the collection of produce but also the purchasing and ordering process for the city's largest produce buyers. Up to twenty-four hours prior to delivery, the "socialist" middleman took orders from Berlin's large grocery stores, regional shops, specialized stores, cafeterias, and factory kitchens. As the belt simplified the production chain, planners hoped it would do the same for groceries and their patrons. In the same summer of 1975, Kiesler advocated to his superior Grüneberg for an expansion in the number of locations where consumers would buy fruits and vegetables from the socialist farms. In one example, he suggested the creation of a model specialty store called the Mecklenburg Village, which would attract consumers with its specialty foods from the northern regions of the GDR.[53] From the heights of the planned economy, the vegetable belt simplified the work of the planner, making it easier for the state to provide the necessary financial support, labor, and machinery for the complicated work of feeding Berlin.[54] On the ground, it proved only marginally successful, as Berliners continued to complain about the freshness of their produce and the lack of variety.

Meanwhile, urban support for garden farming continued to grow amid a series of industrial food crises. In 1975, a pork shortage in Berlin hastened the switch to the Grüneberg Plan. The hot summer and drought of 1976 pushed Honecker to reaffirm the legal place of garden culture. And the "catastrophic winter" of 1979 pushed planners to reevaluate the country's rail, highway, and energy infrastructure, and consequently the future of food production. As Honecker's industrial system faltered, planners looked to the productive advantages of the country's individual gardeners and farmers. Between 1977 and 1981, planners increased the number of tiny socialist middlemen in every community. They established thousands of vegetable and fruit fresh-air markets, erected produce purchasing centers on the back of grocery stores and next to garden colonies, and empowered small butchers to buy up individually fattened livestock. They also instituted new reforms for *Haushaltswirtschaft*, removing limits on the number of livestock that farmers could raise, while also increasing credits, rewards, and fodder subsidies.[55]

EIGENVERSORGUNG

With the ambiguity of the 1970s finally passed, individual production moved prominently into the public eye in 1981. At the Tenth Party Congress in April, the rising cost of energy was on the mind of the Politburo. The main proceedings referred directly to a series of economic crises that had disrupted the economy since the last Party Congress in 1976. The party cited the ten billion marks lost in the severe snowstorms, cold, and power outages of January 1979. The Politburo not only promised to devote more resources to individual production as a way to offset future shortfalls, but also unveiled a plan to remake the delivery of fresh produce around the principle of *Eigenversorgung*, or "self-sufficiency."[56]

The largest obstacle small producers faced was a national infrastructure oriented toward large-scale production of agricultural goods. Thirty years of rural development programs had made farms bigger and more specialized, organizing production around a narrow range of exportable goods such as wheat, barely, potatoes, and pork while leaving the cultivation of fruits and vegetables to either the small number of industrial GPGs or the marginal land of village greens.[57] As one LPG chairmen near Dresden pointed out, "We haven't grown vegetables here in a long time and are no longer capable of such production." In other words, "what was changed in over twenty years of specialization cannot be undone in just a short amount of time."[58]

Such sentiments did not prevent the regime from trying. Like Berlin's vegetable belt, *Eigenversorgung* reduced the overall distances food traveled.[59] Planners targeted inefficient delivery routes, particularly those that moved produce out of the region in which it was produced.[60] In 1983, planners slashed these inter-regional deliveries by 20 percent of their 1981 level.[61] In order to offset the overall drop in industrial produce available by region, planners believed that small producers, like *Haushaltswirtschaft* farmers and VKSK members, would make up the difference. This was not an unreasonable assumption, as small producers raised some 140 kilotons of vegetables in 1982—more than 40 percent of the total inter-regional deliveries for the same year.[62]

The challenge was how to match small-scale garden farming with individual consumers, particularly in the countryside. In response, planners increased the number of produce purchasing (*Aufkauf*) stations for fruits and vegetables, while also increasing the number of retail butchers for pork and poultry. Since vegetable markets for consumers were cheap and required little permanent infrastructure, many regions sought to augment these services and opportunities for small producers and consumers alike. For example, a 1983 report from the Council of Ministers "found complete agreement in the regions, districts, and villages that in order to improve the extent of 'self-sufficiency,' we need to expand direct purchasing opportunities between producers and consumers. Doing so would increase the quality and degree of freshness, while also reducing transportation costs."[63] To that end, the council expanded access for small producers to packaging material, so that they could more easily transport their goods to collection centers and state stores.[64]

FIG 5.2 Gardener selling fruit to produce distributor. August 8, 1976. Photograph courtesy of akg/ddrbildarchiv.de/Heinz Schönfeld.

The region of Cottbus undertook particularly strong measures in their transition to self-sufficiency. In 1983, Cottbus reported back to the Council of Ministers on their progress:

> In all cities and villages the purchasing of vegetables has been assured through the creation of purchasing stores and vegetable markets. In the 601 cities and villages of the region, 1,200 collection and purchasing stations were created for small producers of fruit and vegetables in 1982. These efforts improved the freshness and availability of produce in the summer months especially. Central to these efforts were 365 locations for mobile fruit and vegetable markets. In the district cities of Honestein-Ernstthal, Stollberg, and Hainichen the establishment of weekly markets stood the test in 1982, while also being well received by the citizens.[65]

Weekly produce markets, as well as specialized retail stores (like the fruit, vegetable, and potato stores, or OGS) expanded opportunities for small producers, supplied East Germans with better produce, and built more city-land throughout the country. The expansion of the purchasing (*Aufkauf*) net enabled the incorporation of the hundreds of thousands of small-producers back into the planned economy while reducing the use of fuel for deliveries and transportation. Instead of bringing produce to the consumers, consumers now would have to come to the produce.

EIGENVERSORGUNG IN ACTION

In October 1981, Frank Herold, the freshly minted mayor of Mittelsaida, had the distinct honor of demonstrating for the readers of the *Neue Deutsche Bauernzeitung* (*DBZ*) the lengths his hometown had traveled in the pursuit of self-sufficiency. The village, located in the foothills of the Ore Mountains (Erzgebirge), now planted, raised, and harvested nearly all of its fruits, vegetables, dairy products, and meat. It was a true community effort that had required the cooperation of the local LPG, Mittelsaida's school, the district council, and the local "socialist enterprises".[66] And it perfectly encapsulated, perhaps too well, the ideal form of *Eigenversorgung*."

Nearly every villager, from school children to retirees, had contributed in some way to this program, and all in their free time. Grain farmers from the area's collective farm had planted a quarter hectare of land with cauliflower and red and white cabbage. Students from the polytechnical school watered, weeded, and harvested the same plot, while also planting an herb garden on school grounds. The regional council had planted black and red currant bushes. Members of the VKSK had tended to the plum and cherry trees. The town's retail shop collected and then resold all of this produce to the community.

In addition to garden and vegetable produce, the village also raised a significant number of privately held livestock and animal-based products. Self-sufficiency of this sort did not rely on large barns or feeding sheds. Rather, it was won exclusively through the use of marginal land. After a resurveying of the village, Mittelsaiders found an extra six hectares of space. Of that land, what remained unsuitable for fruit trees and vegetable gardens villagers planted with fodder crops. The extra land allowed the village to raise valuable livestock outside the state's agricultural system. As Mayor Herold proudly told the *DBZ*, "While we planned to raise sixty sheep on this land, we have produced enough fodder for eighty—an unprecedented number. In addition we have contracts to raise ten bulls, fatten another eighteen hogs, and sell as many eggs as possible, as you can see."[67] Self-sufficiency in Mittelsaida relied almost exclusively on the marginal land and a mixture of common and private resources of the village: hillsides and street medians, front and back yards, free time, collective labor, and personally owned animals.

While villages like Mittelsaida attempted to provision their neighbors on public lands, the Tenth Party Congress also extended the *Eigenversorgung* campaign to the collective labor and property of the LPGs. This transition, however, was met with some resistance. Many LPG chairmen refused to devote their precious resources, time, and energy to food production that earned the farm little money. As a result, much of the *Eigenversorgung* work was done without the aid of heavy machinery, chemicals, and highly paid collective farmers. Again, it fell to the marginal members of these communities, retirees and women, to do most of this work. For example, in the LPG T Pröttlin, a retiree named Otto Schieltke planted and harvest by himself almost all of the farm kitchen's vegetables and herbs. Not only had Herr Schieltke produced all of his vegetables on a small one-hectare plot of land next to the entrance to the farm, he used an old horse and small wagon to plow the field and make his deliveries. The work, however thankless, had still managed to transform the tiny plot into a productive bit of land, yielding enough surplus greens to fill five hundred five-liter jars in 1978 alone. In a gendered division of labor that was typical of small production, the farm's women did all the canning.[68]

In addition to vegetables and fruit, livestock also fell under the new self-sufficiency campaign. Pigs in particular could earn farmers a considerable amount of money. Another article in the *DBZ* recounted how a cooperative partnership between three collective farms in the district of Gotha created space for an additional thirty sows, which in turn allowed them to increase their overall output by 15 percent. The complicated process began in the LPG Mühlenberg, which first constructed a shelter in the unused space between two of its existing barns, large enough for the thirty sows. The sows, however, came from the Mühlenbergers partner, the LPG Töttelstädt. Töttelstädt agreed to swap the sows, because it freed up space in their already crowded facilities and made pigs in both farms happier, and thus more likely to gain weight. Finally, the LPG P Schwabhausen completed the transaction, providing the necessary fodder to the Mühlenbergers LPG. All in all, the extra rations combined with the additional space for the sows increased their fertility in both LPGs. The efforts yielded an additional 2,300 juvenile pigs across the entire cooperative, an increase of 15 percent over the quota.[69] The article emphasized not only the benefits and relative ease of these efforts, but also the important economic work of breeding more piglets for the planned economy.

By the early 1980s, the GDR increasingly resembled Migge's rural-urban settlement ideal, albeit with periodic breakdowns. On the periphery of Berlin, a "green" vegetable belt organized suburban production of fresh produce. Within the city limits, older gardens in Köpenick, Pankow, and Treptow were joined by newly established colonies, like Honecker's newest neighbors outside Wandlitz. On the collective farms, farm chairmen and state planners expanded the individual production rights of its members and, soon, agricultural laborers, enlisting school children and retirees in the self-sufficient production of fruit, vegetables, and meat. Taken together, the vast constellation of small agricultural producers raised their output significantly over the decade. In 1980, the VKSK estimated that its members had sold 113,000 tons of fruit, 98,000 tons of vegetables, 1.4 billion eggs, 8,000 tons of rabbit meat, 159 tons of goose meat, and 2,500 tons of honey to state stores. In addition, the VKSK also produced 100,000 nutria pelts and 5 million rabbit pelts, and it designated 872 goats and sheep for export.[70] While the VKSK raised animals, the greatest output in individual meat production came from the countryside. Between 1970 and 1980, annual production of pork by small producers doubled, from 85,000 tons per year to 170,000 tons. This was in addition to an increase of 38 percent increase in beef cattle and a 32 percent in poultry, for a total livestock increase of 74 percent, or a quarter of a million tons of meat in 1980 alone.[71] Even more remarkably, the VKSK estimated that although their membership had topped a million in 1980, more than 5.4 million East Germans participated in some way in the work of small agricultural production.[72]

THE PRICE REFORM OF 1984 AND THE HIDDEN SIDE OF EVERYTHING

As the 1980s advanced, there was a growing sense of division between two separate systems for food production: one industrial, which privileged Western markets, Berlin, and other large cities; and another traditional and decentralized, which was largely oriented toward rural communities. This view, however, was wrong. The industrial system and the traditional system produced each other, and by the 1980s, neither could function without the other. At the overlapping center of these two converging systems were East Germany's garden pigs. This became strikingly clear after the passage of the Agricultural Price Reform in October of 1982 and its implementation in January of 1984.

The reform grew right out of the pork crisis of 1982 (the subject of chapter 7). While the reform was aimed at a variety of individual producers, it targeted individually raised "garden pigs" especially, which were routinely outweighing their industrial counterparts during the crisis year of 1982.[73] The price the state paid for the live animals under the new system would be close to four times the 1982 level—thus increasing the incentives for individual producers to raise more meat. At the same time, the state hoped to even out this price reform by also increasing the "price" of grain and protein fodder for the animals.[74]

Yet planners of the 1984 price reform failed to fully appreciate just *why* the garden pigs of individual producers had become successful. It assumed that every pig was more or less the same. When the plan was announced in January 1983, the price reform induced a curious and unintended shift in the country's supply of pork over the following twelve months that left stores short on pork in the fall of 1983 and oversupplied in January of 1984. It also revealed the true and impressive extent of small-scale, individually produced pork in the GDR.

Garden pigs set themselves apart from their industrial relatives in numerous ways. While the country's industrial pigs were raised in crowded concrete barns, garden pigs lived in cleaner, less stressful environments, free from the epizootic menaces. The garden pig also ate differently. Instead of just grain or chemically enhanced concentrate fodder, it consumed whatever its keeper had at hand, like cooking scraps, leftovers, and even family groceries. The agricultural price reform elided these differences, assuming that garden pigs and industrial pigs were already eating the same things. So while the reform raised the purchase prices for pigs and other livestock, it attempted to even out the effects by also raising the state prices for grain fodder and other industrial inputs. This would have worked out were it not for the fact that East Germany's garden pigs already ate foods reserved for human consumption and did not rely on scarce industrial fodder to grow fat.

Planners had tracked the "wasteful" practice of farmers feeding their livestock food purchased from the grocery store, noting that East Germany's pets and livestock ate almost as well as their owners. An internal report from 1981 revealed that more than 21 percent of all East Germans households kept at least some farm animals, like chickens, pigs, rabbits, goats, cattle, and nutria.[75] When expanded to include pets like dogs, cats, and birds, the percentage of East Germans who lived with an animal was over 40 percent. Furthermore, East Germans fed almost all of these animals a significant portion of their own food—around 10 percent of all the kitchen scraps and 31 percent of all grain produced in East Germany by one estimate. This ratio was even greater for rural households when compared to the urban working class. Although farmers represented a much smaller proportion of the population, they managed to consume seven and a half times the amount of bread as workers, four to six times the amount of milk, and seventeen times the amount of potatoes. These numbers also varied within regions. In rural Schwerin, one village reported around sixty kilograms of rye bread consumed per person per year, while in another the number was three hundred kilograms per person.[76] Wherever planners noted abnormally high rates of bread consumption, individually raised farm animals were sure to follow.

The East German regime's commitment to low food prices had created the conditions for this problem. Unlike in the market economies of the West, the inflationary pressures of the oil crises were not transferred to food. Instead, Honecker and the regime remained committed to this fundamental ideological goal, cheap bread and butter. When the regime introduced the agricultural price reform, however, they failed to consider how the two-tiered price system would affect production. Moreover, they failed to consider the garden pig and its omnivorous nature.

Although the Politburo had approved the price reform plan in the fall of 1982, farmers had to wait a full year, until January 1, 1984, while the state prepared for the transition. During 1983, planners noticed a curious change in the structure of meat production. Until that summer, individual producers had two options for fattening and slaughtering their pigs. The first option followed the traditional system of household production, where farmers fattened and slaughtered the pig entirely outside the planned economy. Generally, when the pig reached market weight, the farmer could send it to a local member of the LPG and have the pig slaughtered—a practice known as *Haushaltschlachtung* (household slaughtering). While *Haushaltschlachtung* pork never entered the planned economy, it could feed an entire family and its neighbors for months.[77] The second option was for farmers to sign a contract with a state slaughterhouse, raise and fatten the pig at home, and then sell the pig back. These contracts came with subsidized fodder rations, the piglets to raise, and a very high purchase price when the pig

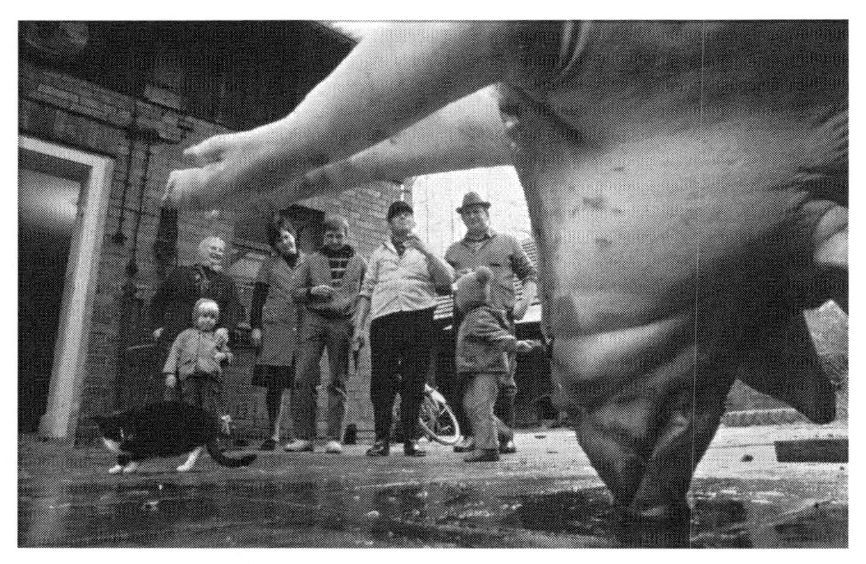

FIG 5.3 Home slaughtering of a pig in Spreewald, October 1981. Photograph courtesy of BPK/Bundesstiftung Aufarbeitung/Harald Schmitt/Art Resource, NY.

reached an ideal weight. Sometimes planners offered additional incentives for every kilogram the pig gained above the target weight.

Farmers began anticipating the coming price reform as early as 1983. In late August, the number of livestock-fattening contracts with individual producers dropped by thirty thousand over the previous year's total. That number continued to increase through the end of the year to one hundred thousand for pigs, and to four thousand for cattle.[78] The number of sow-rearing contracts also fell, possibly endangering the hog population for the coming year. Individual producers, anticipating new prices for their livestock, decided to delay new contracts with the state—foregoing the contractually supplied fodder—and fatten their pigs on their own, outside the planned economy.[79] They knew that the normal savings from the low cost of table scraps would be multiplied in the coming year when the privately fattened pig was sold for the new higher prices the following January.

On January 1, East Germany's slaughterhouses were overwhelmed. In 1983, 10.4 percent of all livestock slaughtered in the GDR came from individual producers; that number rose to nearly 20 percent in the first quarter of 1984. Not only pigs, but also cattle and poultry flooded into state warehouses. Their number rose to 15.5 percent and 26 percent respectively. In the region of Leipzig, the year-to-year increase for pigs from individual producers was even more pronounced, increasing from 12.7 percent to 24 percent. The price reforms affected members of the VKSK as well. While poultry and rabbit meat production rose among rural farmers by 332 percent, it increased even more among small gardeners to 350 percent of the previous year's level. Conversely the number of household slaughtering, e.g. non-socialist production, fell by 20 percent over the previous year's number.[80] Together, East Germany's individual producers and backyard animals flexed their productive might.

A former agricultural planner named Dieter Becker explained, "The LPG farmer spent less in the grocery store for bread, than he received for grain! Individual production was strongly supported by state prices and subsidies, that simply did not hold up to economic calculation."[81] The success of East Germany's garden pig revealed both the limits of industrial livestock production and the curious unintended consequences of mass production of grain and meat. The garden pig connected the industrial and small-scale producer, tying factory fields in Brandenburg to backyards in Berlin.

The 1984 price reform also revealed an interesting intersection with industrial agriculture in Western Europe and the United States—particularly their reliance on subsidies. Indeed, subsidies make the industrial system work, by promoting economies of scale, monocultures, and overabundance. The industrial system only becomes profitable for individual farmers when they expand to a scale large enough to benefit from subsidies. Subsidies in turn affect consumption through the generation of tremendous surpluses, the majority of which cannot be used directly as food. In the United States, surpluses of corn and soy are so great that not even all the livestock in the country can consume them. As a result, these subsidized surpluses find their way into a variety of food and non-food related industries, chemically rearticulated as sweeteners, preservatives, and additives. The extra calories find their ways into everything—into American food and American livestock. From this perspective, corn-based sweeteners bear a close resemblance to the cheap bread East Germans fed to their pigs. The major difference, however, was that the East German regime preferred citizens not feed the subsidized grain surplus to their livestock—that was, until the 1984 price reform.

Garden culture shows another side of state socialism. Instead of just the dirty, gray pollution of state-directed agriculture and industry, garden culture also touched East Germany with shades of green. It was a practice born out of moments of food scarcity, yet it also managed to carry with it some of the aesthetic characteristics of the bourgeois culture that preceded it, despite the objections of communist planners. In retrospect, garden culture provided relatively modest privileges—mostly a little bit of privacy, access to fresh air and green space, and a place to labor for oneself—privileges that made life under state socialism more attractive. To the credit of the regime, it came to recognize gardens as benign and beneficial by the early 1980s, expanding access and allowing citizens to overnight in these second homes. The position of gardens relative to East German economic policies shows how state socialism not only accommodated preexisting cultural institutions, but also incorporated them directly into the cultural identity of the state and its economic plan. Finally, the persistence of garden culture in the GDR shows that rural development in the twentieth century did not just drive people everywhere from the land. While both critics and proponents of industrial agriculture have long assumed that such development unfolded along an urban/rural divide, East Germany remind us of a third-way: the way of the garden pig.

CHAPTER SIX

A PLAGUE OF WILD BOARS

FOR MORE THAN TWENTY YEARS HELMUT ARNDT AND HIS FAMILY HAD SPENT their precious summer vacation camping in the woods around the shores of the Müggelsee, a lake approximately twenty kilometers southeast of Berlin, at the edge of the city's suburban sprawl. Just a short trip beyond the working-class neighborhoods of Friedrichshain, Lichtenberg, and Kaulsdorf, the lake and surrounding areas served as a popular destination for many Berliners. Müggelsee's wooded landscape contrasted sharply not only with the concrete order of East Berlin's city streets to the west and north, but also with the vast fields of the collective farms to the east and south. Both city and rural residents flocked to the Müggelsee to hike the wooded trails or to swim or fish in its cool waters.[1]

The Arndts were among the hundreds of families who would book week-long retreats at the many small vacation villages, garden colonies, and campgrounds spread around the lake district. Of all the activities at Müggelsee, the Arndts most enjoyed their daily outdoor cookouts. Yet their July 1988 vacation, including their cookouts, was ruined. As Arndt complained in an angry letter to the ruling Socialist Unity Party (SED) regime:

> In the past week, we and over one hundred other campers were infested by a herd of wild boars. There were two sows and more than twenty of their offspring, which invaded our campsite twice almost every night.

> They were drawn to the site not by our potatoes or bread, but rather by our meat and sausage products. They rummaged through our coolers with unfailing certainty, to a degree that would surprise even Prof. Meinhard [*sic*].[2]

Professor Meinhard, or, to spell his name correctly, Heinz Meynhardt, was East (and West) Germany's most famous champion of the animal, known in German as *Wildschwein* or *Schwarzwild*. On television, the popular naturalist presented Germans with an image of the wild boar as a social, intelligent, and charismatic animal. It was an image that stood in stark contrast to the monstrous "pests" that had wrecked the Arndt's vacation. This was also not the first time boars had caused problems. In past years, the Arndts, along with many other East German families, petitioned the state for greater animal control around Müggelsee—more traps, more organized hunts. The Arndts even bought an insurance policy for their camping gear, yet the boar menace at Müggelsee only seemed to get worse. Arndt grumbled, "The relaxation value [of the trip] had been zero," since he spent most nights awake, "fearful of what lurked outside."[3]

During the 1980s, such incidents became so commonplace that planners began to refer to it as *die Wildschwein Plage* or the wild boar plague.[4] Campers were not the only ones affected. Grain farmers reported wild boars damaging fields and uprooting crops during the harrow and the harvest. Pig keepers patrolled their facilities to prevent wild interlopers from spreading disease to their domesticated livestock. Weekend gardeners complained that boars destroyed their fences and tore apart their vegetable patches. While the animals had been a problem for decades, they had never seemed quite as bad as they did in 1988.

In contrast, Meynhardt presented the softer side of these notoriously shy animals, at a time when East Germans knew them only by the destruction they left in their wake. Meynhardt was not a trained biologist but a self-taught amateur. He modeled his approach on Jane Goodall's studies of "a wild horde of chimpanzees" by insinuating himself with a group, or sounder, of wild boars in the woods outside Magdeburg (near the town of Burg).[5] From 1974 on, Meynhardt made observations on everything from mating habits and social relations to dietary preferences and habitat range. He wrote books, produced radio reports, and starred in a number of nature films about wild boars over fifteen years. His films were broadcast on TV in both Germanys, but it was in the GDR that Meynhardt used his prominent position to

FIG 6.1 Heinz Meynhardt studies a sounder of wild boars. Courtesy of Hans-Jürgen Ruscyzk, Private Collection.

advocate passionately for the protection of wild boars. On the screen, Meynhardt rubbed the bellies of enormous sows, known as *Schwarzkittel*, fed them corn from the back of his Lada, and followed litters of their striped piglets racing through leafy underbrush. Meynhardt's boars were easily spooked creatures, a stark contrast with the human-habituated monsters that ruined the Arndt's and other families' vacations.

For most Germans, this wild boar "plague" was even more striking, since only forty years earlier, wildlife had all but disappeared from what became the GDR. Centuries of over hunting and habitat destruction had eliminated all sorts of wild animals once considered native to central Europe. Large predators such as wolves, lynxes, and bears, as well as European bison and even tiny voles, were extirpated. The desperate days at the end of the Second World War made the situation critical for even the most resilient of wild fauna, such as deer.[6] Only the Eurasian wild boar (*Sus scrofa*), seemed to shrug off the

Red Army and the hungry civilians, surviving with a large enough population to start growing in number almost as soon as the war came to an end.[7] East German hunters and forestry officials kept their numbers in check until the mid-1960s, allowing for steady but controlled growth. That changed over the next decade. While the numbers are inexact, conservative estimates suggest that the population rose from around 20,000 in 1963 to more than 160,000 by 1989, with a remarkable spike in the middle of the 1970s that doubled the boar count from 52,000 in 1973 to over 100,000 by 1976.[8] In a short time, wild boars went from being rare folkloric creatures to common pests. Many East Germans must have wondered if these animals hadn't just fallen from the sky.

This chapter seeks to explain the surprising ungulate (the evolutionary order that includes pigs, but also deer, sheep, and goats) irruption that inundated the GDR in the 1970s and 80s.[9] It asks: how did this irruption happen, and what does it reveal about environmental change in East Germany? The answer to these questions begins with the regime's halting rural development programs from the 1950s to the 1980s. The expansion of monocultures in the name of the grain-oilseed-livestock complex altered boar habitats, offering a breakfast buffet of industrially designed calories tailored to maximize weight gain and rapid growth in pigs. And what worked for the industrial pig worked for the wild boar.

Rural development, however, was also about the *Landeskultur*, or the entire space controlled by human action and management in the GDR.[10] Dating back to the 1950s, wildlife managers followed this conservation ideology to create layered, multiuse forest and hunting reserves. These reserves, many of which predated the regime by centuries, were not just set aside for hunting and logging, but also overlapped with the property lines of collective and state farms.[11] Planners proudly spoke of these vast reserves, which they believed distinguished the East German system from Western ones. The SED elite loved the reserves as well, promoting hunting to the public as an expression of communist masculinity while simultaneously reserving access to it for the privileged few. No matter the purpose, the regime encouraged planners to make wild game as abundant as possible. Buried within this management scheme, therefore, were the seeds of conflict. Maintaining large wild animal populations for the elite in places with LPGs or campgrounds invited conflict between farmers, campers, animals, and the regime.

The wild boar played a powerful role in this story, subverting the best and worst intentions of East Germans. By refusing to stay in its designated habitat, it neither submitted to the control of wildlife managers nor retreated from the gunshots of hunters. Its desire to roam, root, reproduce, and feed brought it into conflict with people but also made possible its sudden population growth. As we will see, wild boars and domesticated pigs have never been that different. They are two sides of the same coin. Whether pink and fat or bristly and lean, their appearances are merely external expressions of their differing relationships to human beings. They are in fact the same species, *Sus scrofa*. Taking advantage of human decisions to concentrate porcine life and grain supplies in large regions of the GDR, the feral form of *Sus scrofa* flourished. From this perspective, the wild boar plague becomes an expression of the regime's political economy. It was into the managed space of rural East Germany—with its connections to global markets, socialist conservation programs, and elite hunting reserves—that the most "weedy" of all wild ungulates, the wild boar, trod, and promptly did whatever it wanted.[12]

BOAR HABITATS

Meynhardt first encountered wild boars in the fall of 1973, when his friend Rudolf Meseberg—the hunting chairmen of a game reserve outside Magdeburg—invited him to help distribute "distraction fodder" in the forest. Wildlife managers and gamekeepers had long used corn, oats, and other grains to lure hungry game animals away from the fields. Meynhardt enjoyed the feeding work and immediately took note of the boars' intelligence and curiosity. But it wasn't until the following spring that he noticed something remarkable. While helping Meseberg deliver more distraction fodder to a clearing, Meynhardt noticed that "a band of juvenile boars charged out of a thicket of pine trees and immediately, and anxiously, ate the stray kernels of corn. The experienced hunter [Meseberg] was surprised that the little guys displayed no shyness in front of people or the horse," Meynhardt wrote.[13]

The juvenile pigs, however, were not alone. Two mature sows, which initially regarded Meseberg with suspicion from the edge of the woods, eventually came to anticipate the arrival of his horse-drawn cart, waiting attentively along the trail. After a few weeks of seeing him on the wagon with Meseberg, the boars became totally accustomed to Meynhardt's presence. He decided to seize the opportunity to observe.

Over the next three years, Meynhardt spent nearly every day with the sounder, usually driving his Soviet-made Lada station wagon laden with grain

FIG 6.2 Wild boar piglets in the GDR. Courtesy of Hans-Jürgen Ruscyzk, Private Collection.

into the reserve. Once the boars became used to his presence, he was free to sit among them, recording their behavior. Meynhardt recorded feeding, grooming, and mating habits; noted their social hierarchy; located their nests; and tracked their daily movements in and out of the reserve. Worried that his sounder would become too acclimated to human presence—and thus easy targets for hunters—he insisted that other visitors or assistants stay in the car while he moved among his wild boars.[14]

Hunters, Meynhardt soon realized, were not the main threat to the animals. It was the poor condition of East Germany's forests. In particular, he bemoaned the lack of quality fodder in the boars' habitat and cited this deficiency as partially responsible for "poor" wild boar behavior. "The reasons for farm damage are known and illuminating . . . most of our forest reserves do not provide our wild boars with year-round, adequate grub [*Fraß*], acquired through reasonable time and effort."[15] In contrast, the same reserves bordered enormous fields full of cereal grains. It was too tempting for the animals. For Meynhardt, solving the boar problem required habitat restoration, and he used his boars to turn East German attention to the state of their woods.

The end of the Second World War is popularly known as *Stunde Null*, or zero hour, in German history. The people, and soon the Western allies, used the term to suggest that 1945 was the year the country started over, its historic slate wiped clean. In reality, both Germanys could not escape the legacy of what came before. From the practical limitations of denazification to the political constraints of economic reconstruction, the German past remained inseparable from the postwar present. The same was true for forestry, wildlife management, and agricultural development in the GDR.

The year 1945 was a particularly low point for East Germany's forests. The movement of armies and masses of displaced persons left much of the countryside devastated, its rural spaces abandoned, and its wildlife scattered. The ensuing Soviet occupation made a bad situation worse as the Soviets not only took factories, machinery, and railroads, but also clear-cut thousands of hectares of trees as payment for war reparations.[16] The Soviets, however, arrived only in 1945, on the heels of nearly a century of German forestry mismanagement.

Beginning in the early nineteenth century, centralizing states like Prussia introduced "scientific forestry" schemes. Their goal was to remake Germany's forests in the service of industry and global trade. The "scientific" forest favored straight, fast-growing trees like spruce and pine over the mixed species of hardwoods that had dominated European forests into the eighteenth century.[17] Trees in this modernized forest were to serve industrial ends, providing pit props for mines, ties for railroads, cellulose for chemical plants, and pulp for paper mills.[18] Forestry managers planted trees in rectilinear monocultures, first removing brush and the detritus layer. In the process, they permanently reduced the forest's biodiversity. As a result, the forests were severely distressed by the end of the World War I.

The forest decline through the nineteenth century gave rise to a broad conservation movement in the twentieth. During the interwar period, the implications of dying forests, and more broadly the apparent alienation of the people from nature, generated a yearning for *Naturschutz*, or nature protection.[19] This resulted in an eclectic stew of romantic and *völkisch* movements. These ranged from the Nazi's reactionary modernism to the biodynamism of Rudolph Steiner, and from the *Siedlung* (settlement) architecture of Leberecht Migge to the close-to-nature *Dauerwald* forestry of Alfred Möller. As hard as it to imagine followers of Adolf Hitler and Rudolph Steiner in agreement, many did share an inchoate belief that the regeneration of the human spirit was somehow connected to protecting natural spaces for nature's sake, not for economic exploitation or use by the privileged few.[20] These cultural and economic strands formed the discursive foundations for a new form of landscape management and industrial forestry. It was this complicated history that the East German regime quietly inherited after *Stunde Null*.[21]

The GDR's forestry policies prioritized timber production, but also made room for wildlife management and hunting. There was considerable work to be done. Initial estimates of wildlife populations were low at the end of the Second World War. The heavy fighting had devastated habitat and the animals themselves. Many Soviet soldiers had grown up in the countryside, and now, carrying powerful weapons, they went after whatever wildlife remained. With the collapse of the Nazi state and a total absence of hunting restrictions, many soldiers spent their free time stalking red and fallow deer, as well as foxes, martens (relatives of the weasel), and badgers. Paradoxically, the wild boar population grew under these conditions. Soviet soldiers preferred to bag the mature, large-tusked, and notoriously solitary males, inadvertently clearing the way for a baby boom among the females and juvenile males. Disarmed farmers were unable to protect their crops from the surging population. By 1946, Soviet authorities were forced to issue the first postwar hunting regulations to fight the pests.[22]

It was clear from the beginning of the GDR that the country would need to regulate its wild boar population. So in 1949, East Germany's planners placed the responsibility into the hands of forestry managers and hunters.[23] Their new policies reflected the old, ambivalent legacy of prewar industrial forestry, traditional hunting, and German conservation. On one hand, planners wanted to revive the decimated forests and animal populations in much the same way that national propaganda exhorted citizens to "*Bau Auf!*" (rebuild) the industrial economy. On the other, wildlife managers and forestry officials, aware of past mistakes, attempted to adapt the science of ecology and conservation to the priorities of the regime.

This was no easy task. In the 1950s, a small group of academics, scientists, and landscape architects became especially concerned that the rapid pace of reconstruction would endanger natural spaces. They faced strong opposition from within the communist regime, which saw any form of conservation as a remnant of the previous bourgeois/fascist order. In response, the "conservation bloc" adopted a new descriptor, *Landeskultur*, which, as Scott Moranda writes, coopted the economic and political objectives of the new regime while promoting an updated version of conservation.[24]

Landeskultur dated back to the nineteenth-century German romantic tradition that saw an organic link between nationalism, regional identity, and nature. Then, it was broadly referred to as *Kulturlandschaft* (cultural landscape) but in the hands of the conservation bloc and East German planners, *Landeskultur* became the GDR's version of managed-use conservation and "productivity."[25] Under this rubric, the development of forests, wastes, and agriculture for economic production was coterminous with the preservation of nature, especially wildlife. Rather than being in conflict, economics and nature would guide each other. The result would be a more abundant, productive, and protected nature. Planners believed that this approach would lead to the simultaneous expansion and increased productivity of all aspects of the economy, whether factory, farm, forest, or wildlife.[26]

In practice, *Landeskultur* measured the improvement of the forest in terms of raw material produced—tons of lumber, heads of red deer, or kilos of boar meat. The infamous "tonnage ideology" was the gauge. Yet the fulfillment of this productivist logic often left wildlife managers with conflicting tasks. The regime's conservation goals of improving the quality and number of wildlife required managers to restore the natural productivity of the woods. They altered the natural habitat by sowing "grazing areas," inoculating wildlife against parasites, and creating hunting reserves. At the same time, hunting officials were also required to do everything in their power to keep wild game from damaging agricultural and forest production. This led to communal hunts that quickly exceeded regionally determined quotas.[27]

Few members of the elite saw any contradiction in these goals. Hans Stubbe, director of the forestry school in the "Hog City" of Eberswalde, declared in 1988, "Today we know, that hunting and conservation and all subsections of rural culture pose no contradiction to one another, that they are in fact intertwined with one another, that they serve the well-being of all people, and find in nature the source of all regeneration."[28] The resulting consensus was best described by the aphorism *Hege mit der Büchse*—cultivate with the rifle.[29] While forestry managers planted new forest stands and fought over proper harvesting methods, the country's hunters maintained the balance in nature with their rifles, traps, and hunting dogs.[30]

HUNTING AMONG THE ELITES

Forestry and wildlife management doctrine merged seamlessly with an older German aristocratic tradition that valued the woods not only for what they produced but also for who controlled access to them. Hunting especially was embedded in the distribution of power and privilege in the GDR. Communist leaders believed that they were overthrowing the morally depraved traditions of the Junkers, the Kaiser's court, and even the Nazi elite who dearly loved their hunting retreats. After declaring all wildlife the property of the state in the late 1940s, the new communist regime proclaimed that "the hunt belongs to the people." In their propaganda and public declarations, the regime declared hunting and the country's forty thousand hunters as integral to maintaining "natural balances" in the country's forests. In reality, the hunt belonged to the new elite.

Locally, hunting clubs were the preserve of farm directors, production chairmen, and administrators, as the regime restricted membership to ideologically pure, politically connected men. Yet even within this privileged class, there were further layers of stratification. Nationally, none was more blatant than the top leadership of the GDR's ruling Socialist Unity Party (SED), who used their private hunting reserves to indulge in rapacious, ostentatious, and self-serving behavior. From Walter Ulbricht and Wilhelm Pieck to Erich Honecker and Gunter Mittag, all claimed exclusive hunting rights. They justified their self-indulgence on the grounds that they were preserving nature by remaking the East German forest after their own fantasies of a dark ancestral woods filled with "wild" trophies like boars, red deer, and elk. It was the sole domain of brave masculine SED huntsmen.

Hunting was an evocative concept in the GDR, where it was associated with German history and legend. The deep forest is a common setting, from Grimm fairy tales to Wagner operas. For the regime's privileged and elite *Nomenklatura*, hunting harked back to a strong German tradition, ironically with strong aristocratic roots. Game reserves, like Berlin's famous *Tiergarten*, had begun as exclusive preserves for Brandenburg's lords in the seventeenth and eighteenth centuries. The Prussian Junkers and Hohenzollern rulers had also been avid hunters. The Hohenzollern especially played a large role in German hunting history, establishing the large reserve of Schorfheide north of Berlin in the mid-nineteenth century. Over the next century, Schorfheide became the refuge of Germany's most powerful autocrats, including Kaiser Wilhelm, Field Marshall Hindenburg, Adolf Hitler, and Hermann Göring. The rotund Göring transformed the reserve into his personal playground. He ordered the construction of his mansion, Carinhall, between the two largest lakes in Schorfheide. He stocked the grounds with the largest stags and elks he could get from Eastern Europe and wild horses and European bison—his attempt at recreating a Wild West fantasy in north central Brandenburg.[31]

Although the Soviets dynamited Carinhall in 1945, Schorfheide became a beloved retreat for Politburo members. When most of the East German elite moved out of their Berlin residences in the MajakowskiRing in 1960 to the more secure suburban retreat of Wandlitz, they cultivated an association between leadership and huntsman in their public image. While Wandlitz was facetiously known as "Volvograd" among the general population (because of the regime's fondness for Volvo limousines), Politburo members officially referred to it as the *Waldsiedlung*, or forest settlement. Thus, they draped their new seat of power in rustic trappings. The new settlement was halfway between Berlin and Schorfheide, where many Politburo members soon had their own small hunting lodges. Somehow, SED leadership overlooked the strong ties between Schorfheide and the Nazi and Junker association with hunting.[32]

The regime used hunting not only to mark their own power, but also to confer status. A limited number of hunting memberships were distributed to regional clubs, which then could bestow them on local worthies. Membership in these clubs depended almost entirely on personal political connections. Even then, potential members were subject to careful political evaluation. The regime required prospective members to attend classes, the content of which focused almost entirely on citizenship and communism rather than something more practical like marksmanship, safety, or even how to hunt.[33] After an interview, a board of examiners discussed the prospective hunter's ideological suitability. The clubs themselves presented an egalitarian front, but often a single member or faction dominated the association.[34]

The regime made hunters attend ideological training largely because of the danger posed by gun access. If a hunter was going to have a firearm, the regime believed, they had better be loyal to the state. Most hunters in the GDR were not allowed to own guns, since all guns were property of the state. Instead, hunters could borrow rifles from the local *Volkspolizei* unit. Both guns and ammunition were in short supply—which annoyed hunters and severely disrupted the organized "pest" hunts in the 1980s. In total, there were about one hundred hunters granted authority to own their own private weapons. Of those privileged few, the majority were either members of the highest echelons of the SED or senior bureaucrats.[35] Erich Honecker was an extreme example. He owned thirty rifles.[36]

The SED elite not only privileged hunting for themselves, but they also used it as a tool of statecraft, in much the way that golf serves wealthy, powerful men in the United States.[37] During the 1960s and 70s, the regime staged regular *Staatsjagd*, or state hunts, for foreign leaders, who joined members of the Politburo in pursuit of wild boars and stags in Schorfheide. In their book *Jagd und Macht*, Burghard Ciesla and Helmut Suter recount how such excursions sparked the close friendship between Erich Honecker and Leonid Brezhnev in the 1960s. Later, Schorfheide served the new allies as a key location for plotting the palace coups that overthrew their predecessors.[38] The Politburo also used the hunts to convene summits among their allies, even proposing a rabbit hunt in 1985 to serve as the site of an "informal" international summit between Gorbachev and Reagan. Of course, the two leaders would also have been joined by representatives from Czechoslovakia, the Palestinian Liberation Organization, Sweden, Zambia, Afghanistan, North Korea, and Cuba.[39]

No one embodied the power and privilege of hunting more than First Secretary Honecker. He was obsessed with hunting, and he cultivated a public persona as a skilled huntsman, perpetuating the older, aristocratic attitude toward wild animals. Hunting was a proxy for his power, and therefore, the more animals he killed, the more powerful (and more masculine) he appeared. For that reason, Honecker manipulated the animals and environment to suit his needs. He expanded Schorfheide from fifteen thousand hectares in 1970 to over twenty thousand hectares by 1989, reserving most of the land for himself and his second in command, Gunter Mittag.[40] In Schorfheide, both SED leaders were the undisputed rulers. Their technique and total number of hunting trophies reflected their indiscriminate use of power. Mittag, who suffered from diabetes and had had both legs amputated below the knee, hunted sitting down, perched on one of his four customized off-road Land Rovers and Mercedes. Unwilling to be outdone, Honecker also had four imported all-terrain vehicles. Their combined fleet cost DM 1.7 million. These "safari" hunts allowed both men to collect high numbers of trophies. Their bodyguards later remembered that Honecker and Mittag routinely bagged dozens of elk, deer, and wild boar on every expedition. Between 1979 and 1985, the two men shot 243 stags; 126 by Honecker and 117 by Mittag.[41]

Beyond the borders of the private estates, the hunting of wild animals served a different purpose for ordinary people. Hunting became a means to maintain a healthy environment. Planners continued to insist, well into the 1980s, that under the principles of *Landeskultur*, hunting served the goals of conservation *and* rural development. Landscape planners could manage natural spaces as layered, multiuse preserves. In the state hunting reserves, many of which predated the regime by centuries, *Landeskultur* did not reserve access for just hunters. Planners intentionally drew hunting boundaries that overlapped with the property lines of collective and state farms, forests, factories, and even recreational preserves.[42] Natural and human-made features formed their borders. Hunting—cultivation with the rifle—served the interests of multiple economic and recreational users in *Landeskultur*. It fulfilled the desire of forestry officials to promote the health of forests and wildlife, it allowed farmers and agricultural planners to limit the crop damage by wild animals, and it gave "the people" the chance to pursue trophies. Planners proudly referred to these vast, unenclosed hunting grounds as "large-scale" preserves and pointed to their management techniques as examples of the superiority of the East German system.[43]

Wild boars, however, continued to undermine their best laid plans. The overlapping, multilayered reserves and economic zones of *Landeskultur* brought the animals into greater conflict with farmers, gardeners, and weekend travelers. As we've seen, the 1976 Grüneberg Plan had filled the countryside with more potential boar fodder than ever before. In this way, the wild boar "plague" was connected to the global transformation of agriculture. It revealed another aspect of the Grüneberg revolution, beyond the regime's debt ledgers, its flows of grain and pork, or even its epidemic of mossy giants. But what did that change look like to a wild boar?

LANDSCAPES OF DEVELOPMENT

On May 29, 1973, the American satellite Landsat 1 passed 917 kilometers above the GDR. Completing one of its fourteen daily orbits of the planet from the northern polar ice cap to the South Pole, Landsat 1 took 175 kilometer-wide photographs of the country using a new technology known as a multispectral scanner.[44] That day, as the camera recorded the distinct bands of electromagnetic radiation refracting off the surface of earth, the satellite captured the East German countryside on the eve of the implementation of the Grüneberg Plan. For the next seventeen years, Landsat satellites continued to record changes in the land across the GDR. A Landsat image of the same area taken fifteen years after the first, on May 8, 1988, provides a standard for comparison. Together, the two images span the implementation of the Grüneberg plan, revealing the ways in which it remade the countryside.

The images cover an area bounded by Poland on the east, separated from the GDR by the Oder-Neisse rivers. The large forests and lake districts of Waren-Muritz and the Ueckermark run across the bottom left quadrant; at the top left of the image are the agricultural towns of New Brandenburg and Anklam, while the top-right side includes the Oder Lagoon, which is separated from the Baltic sea by the island of Usedom. The bottom edge of the image is approximately one hundred kilometers north of Berlin. The images are classified by land usage, meaning that the lightest shade of gray is an amalgamation of all cultivated and fallow farmland. The black areas are forest cover. The white is water and the middle shade of gray represents urban spaces. This is not the entire GDR, but these satellite images cover the large central agricultural plain, which contained some of the highest concentrations of wild boars. Landsat provides the perfect vantage point to see how both planners and wild boars saw the landscape evolve.

The first thing to notice is the increase in area covered by forest. Whereas the 1973 multispectral scanner showed the agricultural plain between the shores of the Baltic and the central forests to be largely devoid of trees, the 1988 image shows a remarkable reforestation. From 1973 to 1988, farmers combined medium-sized parcels into enormous new ones, leaving marginal and "uneconomic" spaces in between to become reforested. While remote sensing analysis is a powerful tool, it is only one method among many that helps us determine the cause of the wild boar irruption. Still, it provides a clear visual representation of the ways in which forests and farmland developed together, making new boar habitat in the process.

These changes were not lost on the chief wild boar expert, Lutz Briedermann. The onetime director of the venerable Forestry Institute in Eberswalde, Briedermann was well read in the growing literature of ecology from all over the world, including the works of Aldo Leopold, one of the founders of modern American environmentalism, who had traveled to Germany in the 1930s to study its forestry and wildlife management practices.[45] Briedermann borrowed the concept of "edge spaces" from Leopold to work out formulas for measuring wildlife density.[46] In the 1970s and early 1980s, Briedermann created the national guidelines for optimal population densities of wild boars and other game. His calculations relied on two primary factors: the percentage of hardwoods in the total forest area, and the total length of edge space between farms and forest. The more hardwoods a forest had and the greater

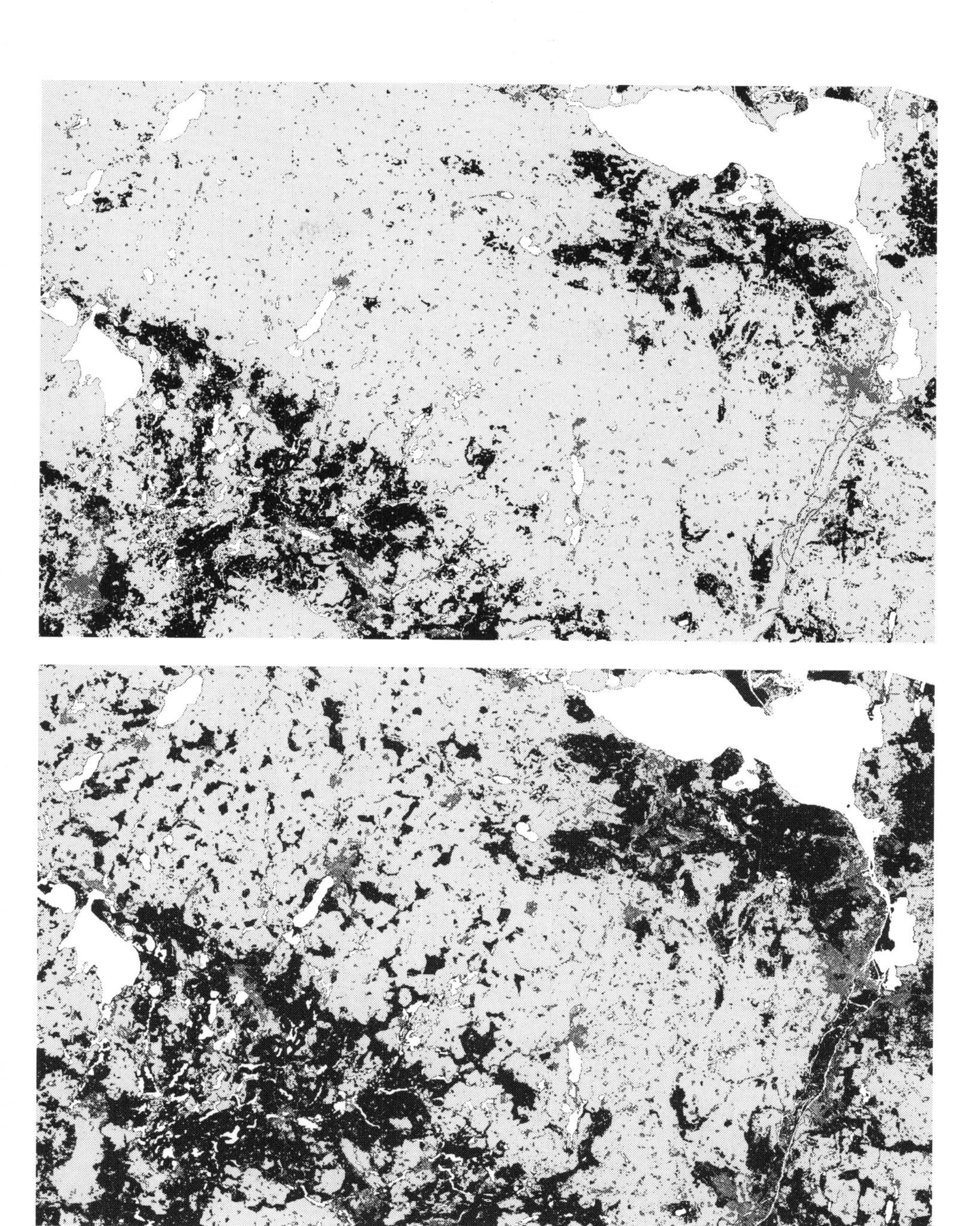

FIG 6.3 These two maps use "false colors" to make the differences in land usage more visible. To summarize, light gray represents farmland, dark gray is urban space, black is forest cover, and white is water. (*Top*) Northeast GDR. Landsat 1 MSS, classified, May 29, 1973. (*Bottom*) Northeast GDR, Landsat 4 TM, classified, May 8, 1988.

exposure of the forest's edge to cropland, the higher the density of wild boars in the area, according to Briedermann's formula.[47]

Briedermann's guidelines addressed the question of whether or not wild boars, and the damage they caused, could be controlled. Hoping to find a way to keep the animals safe from angry farmers, he ran several studies on the relationship between crop damage, wild boars, and the availability of "distraction fodder." In a 1980 experiment, he argued that wild boars in fact preferred to remain within their own established range (approximately one hundred square kilometers), if quality food sources were made available. Through tagging and recapturing of 1,762 wild boars in the Wildlife Research Center in Wierzen, which he had stocked with corn and other grains, the study showed that 91 percent of the recaptured boars and 95 percent of the sows stayed within a ten-kilometer area.[48] In other words, wild boars could be successfully decoyed and controlled. Briedermann wasn't alone in his thinking; Heinz Meynhardt, the TV naturalist, voiced the same conclusion in his books and television appearances. "I cannot endorse the conventional wisdom on wild boars that they restlessly wander through their reserves, today here and tomorrow there," wrote Meynhardt. "Quite the opposite, if they remain undisturbed and there is enough to eat, the wild boar is territorially true. . . . The more beloved a reserve is, the more unlikely they are to leave."[49]

Both Meynhardt and Briedermann blamed boar "misbehavior" on failures to provide enough distraction fodder, as well as other "disturbances." While Briedermann maintained that boars could remain in their regular reserves, he had to contend with evidence that they didn't. In the same 1980 study, his researchers only managed to recapture 452 of the 1,700 animals they had tagged. So while the study claimed that 90 percent or more of the animals stayed within the ten-kilometer area, more than two thirds of the test subjects were gone. Furthermore, of those recaptured, eight sows had been found more than fifty kilometers away.[50] To account for the missing animals, Briedermann blamed human disturbance of boar habitat, and particularly hunters and their dogs moving too close to the relatively small forest stands the boars called home. According to the naturalists, wild boars only left the preserves when pushed by humans rather than pulled by greener fields elsewhere.

Both Briedermann and Meynhardt failed to consider how wild boars were adapting to a changing environment. They proceeded from the assumption that the wild boar's true habitat was in a deep primeval forest, much like Erich Honecker's huntsman fantasy in Schorfheide. Wild boars refused to stay put, they argued, largely due to the "unnatural" state of East Germany's woods. While Briedermann continued to develop management guidelines that incorporated the concept of edge spaces into the size and shape of the GDR's reserves, the wild boars continued to subvert his expectations. In doing so, they revealed the extent to which rural spaces had been transformed to the advantage of the wild boar. In short, wild boars didn't live just in the forest anymore.

Both historical evidence and the Landsat maps provide clues. First, recent scientific studies on the relationship between agriculture, forests, and *Sus scrofa* have shown that wild boars are especially adept at utilizing habitat made up of both small and medium forests stands (less than one hundred hectares) alongside large areas of agricultural cultivation. The 1988 Landsat image shows an agricultural plain divided by a "wildlife highway" connecting the Baltic coast to the large forested areas with small and intermediate forest stands on either side. Ecologists have shown that not just individual forest stands influence the growth of wild boar populations, but rather the proximity of small and large forest stands to one another. Fragmented forest and edge habitat alone cannot support large populations. When they remain within one hundred kilometers of a larger forest stands (anything one thousand hectares or more), their carrying capacities increase significantly.[51] By 1988, this type of habitat dominated the landscape of northeast East Germany.

Collective farmers in the 1980s frequently worried about boars destroying their crops or bringing disease to their animals. On July 28, 1981, the chairman of the LPG P Saletal wrote a letter to Bruno Lietz, head of the Politburo's Department of Agriculture, demanding help in fighting an ungulate infestation. "On behalf of our party organization we ask for your help, in order to avoid excessive wild animal damage to our potatoes, grain, and corn crops. Day after day the damage grows worse, prompting us to bring our problems to the highest authority."[52] The regime's remuneration payments to farmers for crop damage more than doubled in the ten years between 1979 and 1989. Since the enormous size of collectivized fields made careful inspect difficult, the total amount of crop damage was certainly much larger.[53] Wild boars also endangered healthy pig herds by spreading disease. According to biologists, wild animals can act as reservoirs for livestock diseases, including tuberculosis, trichinellosis, hog cholera, and Aujeszky's disease.[54] During the 1980s, the regime took increasingly drastic measures to prevent transmission, like emergency cullings of wild populations.

Despite the growing evidence of the changing structure of wild boar habitat, many people, from the top of the regime to the general public, shared Briedermann's and Meynhardt's belief that the true wild boar habitat was in a primordial forest. When boars acted out, it was because they were out of place—in a worker's garden, a collective farmer's field, or a family's campsite. The wild boar problem, however, was not a matter of wild boars leaving their reserves—the whole country had become their reserve. A look back at the natural history of *Sus scrofa* reveals that wild boars have never preferred one type of habitat over another. Using their skills for adaptation, they have modified their bodies and behaviors to suit whatever environment has worked best for them, whether it is medieval forests or industrial monocrops.

WILD BOAR HISTORY

The Eurasian wild boar is the common ancestor of every single breed of domesticated pig in the world, moving in and out of a domestic relationship with humans many times over the past nine thousand years.[55] It was only during the agricultural and capitalist revolutions of the eighteenth century that farmers moved pigs into the controlled environment of the barn and sty. Total environmental control yielded modern breeds, and marked a fork in the evolutionary road for *Sus scrofa domesticus*. The fat pink or white pigs of our imaginations, from Wilbur to Babe, are relatively recent phenomena.

For thousands of years prior to the agricultural revolution, *Sus scrofa*'s relationship to humans covered a spectrum of possibilities, including fully feral, semi-feral, and domesticated. In the ancient Roman landscape, swineherds rarely kept their pigs in pens for long, but rather let them to roam about to graze and forage on their own. Early "domesticated" pigs were also tough enough and social enough to drive off most predators.[56] Since these pigs spent most of their lives foraging for food and fending off predators, premodern European pigs were built for survival, thinner and more muscular than their industrial descendants. In early modern Europe, semi-feral pig keeping remained a major component in subsistence peasant agriculture. The practice, however, declined with the enclosure of common lands and the depletion of European forests in the eighteenth century, but in small communities, for example, among the Lipovaners in Romania, letting pigs forage for themselves persisted well into the twentieth century.[57]

The long history of domestication has never blocked regression. As we have seen, barnyard pigs can revert to their wild form in a single generation, given the right conditions. If a young pig is exposed to hardship shortly after birth, and a series of transformations take hold—its skull and legs will grow longer, its ears will stand erect, and bristly hairs and a spiked mane will burst from the crest of its skull to its tail. In just one generation, this formerly domesticated pig will appear as aggressive and wild as its ancestor *Sus scrofa*.[58] Technically, these pigs are known as feral hogs and are rather hard to tell apart from their truly wild relatives. Meynhardt related a story about visiting the famous Danubian pigs of the Lipovans in Romania. These pigs lived almost the entire year in the brackish waters of the delta, placed there by their Russian-descended swineherds. In December, the Lipovans would take to boats to round up their pigs, using a little grain as bait and whistles to which the pigs had been conditioned to respond. With winter coming, Meynhardt watched the hungry animals swim for kilometers, trailing their owners' boats back to the village. Were it not for the whistle call, Meynhardt noted, he would never have been able to tell the difference between Lipowaner pigs and the feral sounders that moved through the woods.[59]

For *Sus scrofa*, phenotype (physical appearance) and genotype (genetic makeup) correlate less and less the closer one looks. Zoologists have long tried with limited success to establish taxonomical checklists using phenotypical markers such as split tips on the guard hairs or striped patterns in juvenile pelts. More recently, geneticists have attempted with mixed results to sort alleles in chromosomes. Attention to bone structure has also failed to yield definitive results, as environmental conditions such as climate and diet have variable effects on the size and shape of skulls and limbs, leading biologists to mistake feral hogs for *Sus scrofa* more often than not.[60] There are few absolute criteria for identifying wild boars.

This has not stopped some from trying to enforce a strict line between feral hogs and domesticates. A study led by biologist Alain Frantz looked at the resurgence of wild boars in England, some seven hundred years after the animal's extinction. The study tried to determine the percentage of "true wild boar" in the feral hog population through mitochondrial DNA tests.[61] Instead of a "native" wild boar, they found traces of Chinese hogs, a range of European domestic breeds, and ostensibly some "pure" *Sus scrofa*. Frantz and his colleagues, citing the genetic evidence as proof of the invasive and unnatural quality of these animals, failed to consider the long history of human-pig relations, in which pigs and people have lived together in varying degrees of cooperation and dependency, of wildness and domestication. Instead, they invented an arbitrary true wild boar against which to measure a spoiled, anthropogenic variant—the hybrid wild boar. This "true" wild boar, however, is a myth. Just as the domesticated pig carries the wild boar in its genetic code, so too does the wild boar carry the history of domestication in its. A more accurate definition of the wild boar would account for the ways in which *Sus scrofa* has moved in and out of direct human control for hundreds of years, mixing, borrowing, and sharing its DNA with its "improved" or even "vigorous" cousins. Keeping this insight in mind, East Germany's ungulate irruption takes on new significance. In truth, it could trace its origins to both inside and outside the collective farm.

Raising tens of thousands of animals in industrial confinement while devoting as much arable land as possible for grain production meant that there were more pigs in East Germany than had ever existed before—some eleven to thirteen million in the GDR and hundreds of millions more across Europe. Industrial hogs in the GDR were the descendants of the first scientific breeds, created by the emergent scientific culture and economic relations of industrializing Europe in the nineteenth century. These "improved" breeds, all the old unimproved peasant stocks, and all the wild pigs lurking in the forest carry the same *Sus scrufa* genome. Genomic studies have failed to come up with the biblical Adam of *Sus scrufa*. The industrial pigs of Eberswalde and the wild boar watching them from the edge of the woods are in fact the same creature. Like all their ancestors, the industrial pig possessed the ability to escape and become feral. The explosive expansion of hog farming, rapid reforestation, and the ready supply of monoculture fodder made the GDR a paradise for pigs escaping domestication. Agricultural policy in the 1980s likely increased the opportunities for feral and domestic populations to meet, mix, and exchange DNA. In this way, East Germany's pigs acted in a way they had always done, and yet they were made "modern" by the state's pursuit of grain, oilseed, and livestock production. The irruption heralded not the return of an ahistorical, folkloric species but a new configuration of millennia-old human-pig relationships.

To see the post-Grüneberg landscape through the wild boar's eyes (and its other keen senses) is to see East Germany as a habitat remade for the wild boar. Spreading out of its early twentieth-century ranges, where clear cutting, industrial planting, and over hunting has exhausted the forest, *Sus scrofa* roamed wherever it found food plentiful and hiding spaces easy. Meynhardt observed one sounder easily evading hunters by utilizing the edge spaces that led through farm, field, and forest, and back again. "Between two larger wild boar ranges, which were separated by a four-kilometer-wide field and small stream, an animal trail ran, which several older hunters claimed had been there 'for ages.' . . . In March and April I often sat near this area . . . and observed a large sounder of wild boars moving between these reserves twice a day. Their range in fact was *both* reserves."[62]

In their journeys and their conflict with people, wild boars by the thousands revealed the temporal and spatial contours of industrial development. The sudden rise in their numbers coincided with the East German turn toward global trade in the 1970s, which pulled the rural landscape into the web of capitalist ecology. In the era of global development, there is no industrial pig without the wild boar. Crop destruction, livestock epidemics, and wrecked camping trips were the inverted costs of extensive grain monocultures, confined feedlot operations, and a consumer society built around cheap food. While East Germans worked to keep domesticated pigs and wild boars separate—as the former served the ends of maximum grain/meat production and the latter endangered them—the multiple bodies of *Sus scrofa* challenged this constructed boundary. They evaded quotas, fences, and rifles that were supposed to hold the line between wildness and domestication. In doing so, *Sus scrofa* emerged as an agent of agricultural development and its byproduct. Feral and tame, agent and product, the dual nature of East Germany's pigs connects environmental change and political economy on both sides of Iron Curtain.

CONCLUSION

As East German agriculture expanded in scale and intensity in the late 1980s, the country's cultivated areas encroached on woods and marginal spaces, bringing wild animals, farmers, and, most surprisingly, Berliners into conflict with boars. City residents reported boars roaming their streets. In 1983, the municipal authority of Berlin issued a backhanded acknowledgment of this fact in an ordinance regulating hunting. Wildlife in the city, the government declared, had reached a "good and widespread" density while in some areas even rising to levels "higher than what was permissible." In December 1984, the authorities posted a new regulation for the besieged Berlin suburbs. "In response to the repeated petitions of citizens of the capital city about the increased amount of wild life damage to allotments, gardens, and other areas," the city's upland forestry chief abrogated the existing hunting regulations, in effect declaring open season on all large ungulates.[63] In 1987, the regime formally abandoned any pretensions to "cultivate with the rifle." Instead, they ordered the indiscriminate culling of all wild boars with traps, poisons, and guns in a desperate attempt to bring the plague under control.[64] Although couched in the strained language of socialist bureaucracy, the regulation acknowledged a limit to the density of wildlife in the GDR, especially when that density consumed the tomatoes, lettuce, and snap peas of furious East Berliners.

The wild boar problem now seems oddly familiar, presaging similar phenomena that have since occurred in most developed countries. Around the world, governments are contending with wild ungulate "bombs" in their backyards. They raise important questions about the still vexing relationship between nature and human culture in this era of the Anthropocene. The presence of wild boar in the world's most developed cities is indicative of broader and more common environmental shifts, like the rise of the grain-oilseed-livestock complex—transformations that began earlier in the GDR. Perhaps even more interestingly, the persistence of ungulate bombs is an ongoing legacy of the GDR in the world today. This is the power of *Sus scrofa* in history. It extends the temporal and spatial limits of the East German experiment to times and places where the GDR regime never held sway. Pig bodies and pig behaviors are as much a part of human history as any farmer, hunter, or TV naturalist.

CHAPTER SEVEN

THE IRON LAW OF EXPORTS

WERNER FELFE ARRIVED EARLY AT THE OFFICE ON THE FIRST WORKDAY OF the New Year, Monday, January 4, 1982. Located in Berlin's central district of Mitte, his office in the "House on Wederschen Markt" had housed the Foreign Office during the Weimar Republic and the Third Reich. In 1959, it had been handed over to the leadership of the Socialist Unity Party (SED). By the 1980s, the building was home to the office of the GDR's chief agricultural planner. This was Felfe, who had been in the role since the spring of 1981, following the death of his predecessor, Gerhard Grüneberg. After a tough year, Felfe had reason to be optimistic. The weather on January 1 had been surprisingly mild. Spring-like temperatures rose into the mid-40s from the Baltic Coast to the Ore Mountains, bringing an early thaw.[1] It seemed as if nature might give the GDR a break that winter.

Felfe alas found bad news awaiting him. Reports on his desk described a complicated and multifaceted set of problems affecting the country's cattle and pig farms. A predicted fodder shortage had turned out to be much greater than expected. While planners had projected a shortfall of 350 kilotons of grain, experts were reporting an actual deficit of seven hundred to eight hundred kilotons—more than twice the original estimate. In addition, there was a 182-kiloton deficit in protein fodder from fishmeal and grain silage. Without substitute fodder, Felfe calculated that the GDR would suffer an immediate "production deficit" of more than 400 kilotons of milk as well as 140 kilotons of meat—the equivalent of 1.4 million pigs.[2]

Then the pigs started dying. The following weekend, farmers in the district of Güstrow in the northern region of Schwerin reported several cases of transmissible gastroenteritis (TGE) to the Office of Veterinary Matters within the Ministry of Agriculture. The outbreak began on January 9 in a herd of 1,600 grower hogs, or *Mastschweine*, on an LPG (cooperative farm) in Reimershagen. By January 11, it had spread to three other collective farms in the area, potentially infecting some fifteen thousand pigs. While TGE was a mild infection for adult pigs, it was deadly for piglets, particularly those less than three weeks old. The main symptom, diarrhea, was fatal to piglets.[3] Against the background of the unexpected fodder shortage, TGE was especially ill timed for Felfe, who was already worried about the month's projections. Felfe ordered the affected herds immediately immunized and the infected animals separated from the broader livestock population. Planners also placed farmers across the region on alert for the disease.[4]

Before the ministry could snuff out the TGE outbreak, farmers reported another hog illness. On January 14, an LPG chairman in the district of Teterow, in the region of Neubrandenburg, reported a case of hog cholera (also known as classical schweinepest) in their LPG Gnoien, which had about two thousand pigs. Hog cholera was another devastating disease. The symptoms included general exhaustion, sleepiness, and a characteristically high fever. An infected sow could pass it to her piglets in utero, which in turn could lead to still births or mummification of the fetuses.[5] If the piglets survived their birth, they tended to shiver, look chilled, and huddle in groups.[6] A hog cholera epidemic would have been a disaster for the pig population. Though only two sows had died of the disease, planners from the regional council—the deputy chairmen of agriculture for Neubrandenburg—ordered the entire herd in Gnoien liquidated, at a total loss of 1.39 million East German marks. The farm would have to wait until May to begin rebuilding its stock.[7]

The rash of epizootics was troubling, as it was coupled with a spike in sow infertility and declining piglet numbers. East German sows had given birth to 15,800 fewer litters over the last four months of 1981 than they had in the same period in 1980. The drop in sow fertility cost the country's farms at least eighty thousand productive feeder hogs. The disappearance of the country's fattest hogs, however, put the GDR into a state of emergency. On January 22, the Ministry of Agriculture reported to Felfe that the overall number of the country's two heaviest weight classes—80–100 kg and 100 kg plus—had shrunk by nearly seventy thousand animals, or 29 percent of the previous year's finishing hog population. This decline represented a loss of thirty kilotons of pork.[8] Even more shocking, the finishing hog population fell by half a million in the month between New Year's Eve and February 1.[9] The fodder shortages had been troubling and the disease outbreaks worrying, but now Felfe had to contend with the chance of total collapse of the country's animal farms.

The East German regime in 1982 stared into an economic abyss. A calamitous die off of 1.7 million industrial pigs threw the planned economy into chaos. What started as a shortfall in industrial grain fodder in January turned into shortages in pork, butter, and milk by April, and then coffee, toilet paper, and potatoes in October. All of the problems that had plagued industrial agriculture since the 1970s, from the inability to produce enough grain fodder to the poor condition of the country's livestock facilities, converged in 1982. While industrial pigs declined, wild boars and garden pigs were flourishing in the GDR's marginal spaces. Planners now worried that wild boars had reached such numbers that they posed a threat to the already troubled health of industrial pigs. With state-grown meat supplies in peril, the record number of garden pigs appeared as a glaring rebuke to economic planners and to the state system.

At that point, the regime's deep internecine conflicts over agricultural development erupted anew. The old struggles over the Grüneberg Plan were refought. Once again, the regime's power players squabbled over the issue of Western grain imports and whether the GDR should prioritize its Western trade partners over its domestic population. The debate was existential. The question was whether the regime had more to fear from their Western creditors or from their own population. Willi Stoph, chairman of the Council of Ministers, put the pro-export camp's position most forcefully in a speech in May of 1982. Even as the country was in the grip of shortages, Stoph lambasted economic ministers who questioned what he called the "iron law of exports."[10] Stoph believed that for the state to survive, its Western debt needed to be serviced. That meant pork exports must continue as contracted. As the year progressed, East Germans began to resist. In the early stages of the shortages, they complained loudly in grocery store lines and factory cafeterias. When the crisis did not abate, they wrote petitions (*Eingaben*) to the regime. The more the unrest grew, the more nervous the government became. Believing they were on the brink of a widescale uprising, the regime turned to the West for a billion-mark bailout, irreversibly hitching the fate of the East German economy to the capricious demands of West German banks.

THE RECKONING OF 1981: SOLIDARITY AND DEBT

The pork crisis of 1982 did not appear out of nowhere. It came on the heels of a heightened moment of Cold War conflict. The early 1980s marked a new and darker era for the GDR as the sunlight of détente faded. The year 1981 had been particularly unsettling. To the East in Poland, the massive protests mounted by the Solidarity movement were threatening to bring down the entire communist regime. In the West, economic upheavals unleashed by the oil shocks of the 1970s were boomeranging back upon the GDR, as the era of cheap money and cheap oil ended. The regime had to borrow increasingly larger sums from foreign banks to pay the accruing interest on the loans it had taken to launch the generous social spending, economic development, and the industrialization of agriculture.

To make matters worse, in August 1981, the Soviets informed their East German counterparts of another cut to oil exports, reducing the amount by two million tons annually. The East German regime fumed, as the cutbacks undermined the country's oil refining industry, which had become a lucrative source of Western currency. General Secretary Erich Honecker did not react well. In a letter to his friend and patron Leonid Brezhnev, he asked for a reprieve, claiming that the reduction in oil deliveries "undermines the foundation of the GDR's existence."[11]

Unnerved by Honecker's negative reaction, in September 1981, the Soviets sent the chief of Gosplan (the Soviet Planning Administration), Nikolai Baibikov, to placate the East Germans. Instead, Baibikov was berated by Gerhard Schürer, the East German chief of the State Planning Commission (SPK), who warned, "Imperialism stands right at the door of our house with its hate on three television channels. Now we have the counterrevolution in Poland at our backs. If the stability were endangered here, it could not be restored with 3.1 million tons of fuel." His protests were ultimately useless. The natural wealth in oil and minerals of the Soviet Union were already propping up the entire Eastern Bloc, including crisis-ridden Poland, as well as the USSR's own domestic economy.[12] The Soviets could do the East Germans no more favors.

The pressure from Western capital and Eastern unrest came to a head during a three-day official visit, December 11–13, 1981, with the West German chancellor, Helmut Schmidt. Fittingly, pigs played a part. Prior to Schmidt's arrival, Western creditors had started to express their unease over their loans to the GDR in the form of higher interest rates.[13] For much of the previous two years, Günter Mittag and his coterie had scoured the globe for new lenders in France, Austria, Britain, and Kuwait. It was all to no avail. Global investors had quietly made the Eastern Bloc a no-go zone for capital after the Solidarity uprising.[14]

The 1981 summit took place at Honecker's wild boar hunting lodge, *Hubertusstock*, in Schorfheide. Here Honecker reportedly asserted to Schmidt that the East German industrial economy was the seventh largest in the world, while at the same pleading for financial assistance from the West.[15] The two leaders also argued over the fate of Poland and Solidarity. Honecker himself had been secretly pressing his Warsaw Pact allies to suppress Solidarity. On the last morning of the visit—December 13—both leaders were surprised to learn that General Jaruzelski had declared martial law in Poland.[16] The two

FIG 7.1 At the train station in Güstrow (GDR), East Germany's Erich Honecker hands a parting bonbon to his counterpart, West Germany's Helmut Schmidt, before the chancellor's departure after a three-day summit, December 11, 1981. Photo courtesy of BPK/Bundesstiftung Aufarbeitung/Harald Schmitt/Art Resource, NY.

men muddled their way through the rest of the summit (which included a two-hour drive together through the snow to the northwestern city of Güstrow), before Schmidt departed on a train for the West.

Over the next week, the Polish military under Jaruzelski crushed the trade union, arresting its leaders and putting an end to the rolling strikes that paralyzed the country. Despite the efficacy of Jaruzelski's crackdown, the situation continued to worry the members of the GDR Politburo. While issues like the control of civil society, independent union representation, and Poland's relationship to the Eastern Bloc were at the heart of Solidarity, the East German leaders knew that an increase in the price of bread had lit the Polish tinderbox. It was within this context, that Honecker and his most trusted advisor, Mittag, brokered a trade deal intended to hold back the forces of Western capital and buttress domestic production.

For Honecker and the tight cadre around him, lowering East German debt, maintaining political stability, and delivering cheap food were essential to their survival. These issues were also deeply interconnected. While the regime wrestled with a new geopolitical order and global economy, the leadership put their faith in the country's agricultural sector to provide a way forward. Many planners, particularly those who represented the interests of the SPK, believed that the specialized collective farms were finally becoming self-sufficient (i.e., no longer dependent on Western imports), which would enable the GDR to ride into a golden future as an "export land." In October 1981, the regime took off the training wheels. Mittag brokered a pork trade with the minister-president of Bavaria, Franz-Josef Strauss, while also consulting with the SPK's director Schürer, who was designing the economic plan for 1982.[17] Together, Mittag and Schürer agreed to lower the country's debt load by a onetime increase in pork exports brought about by a drastic downsizing of the national pig herd. A smaller herd would put East German agriculture on the path to grain independence with steady but steep reductions in foreign grain fodder imports—cutting back about a million tons a year for each of the next three years.[18]

Thus, the Mittag and Schürer plan would simultaneously reduce in the country's hog population and its Western grain imports. This "adjustment" would yield somewhere between 1 and 1.5 million hogs for the 1982 export plan for sale to West Germany and other non-socialist economies (*Nichtsozialistischen Wirtschaftsgebiet*, or NSW) for valuable foreign currency, also known as *Valuta*.[19] Schürer's plan was a master stroke of budgetary jujitsu, with one blow flipping a chronic deficit into a bonus. The logic was impeccable. The proportional number of hogs culled would no longer need Western grain to eat while then selling these surplus animals to West Germany would yield a handsome onetime profit for the GDR. Unfortunately for Mittag and Schürer, their perfect plan crashed into the reality of January's livestock numbers, setting off a rapid retrenchment of the industrial pig population.

JANUARY AND FEBRUARY: MISSING PIGS

In January 1982, the Politburo raced to keep up with the struggles of the country's farmers and their pigs as the sudden cutback in grain imports hit home on the collective farms. The Central Administration for Statistics later estimated that the total number of pigs fell by 1.2 million between October 1981 and March 1982.[20] A veterinary report estimated that pig deaths had climbed to more than 150 percent of the previous year's numbers in some parts of the country, while several livestock LPG's reported they had no *Mastschweine*, or finisher hogs over eighty kilograms—a catastrophic figure.[21] In the four most northern regions of Rostock, Schwerin, Neubrandenburg, and Magdeburg, the population and weight declines had likely led to a drop in the pork supply of fifty-three thousand kilotons of meat, the equivalent of 530,000 hogs.[22]

Studying the statistics in late January, Werner Felfe could see two sizable gaps on either end of the pig production chain, as losses grew in the number of piglets and the shortage of *Mastschweine*. The disruptions presented a conundrum to Felfe and his coworkers. Plan numbers were often descriptive, hiding more than they revealed. The statistics could not answer critical question such as what was making pigs sick. Instead, the reports led only to other complex questions: Was illness the result of infectious agents or malnourishment? Are pigs dying or simply putting on weight at a slower than expected rate?

Finding an answer was not easy. From the top-down perspective of the planned economy, it was difficult to tell the difference between dead pigs and so-called *Verluste* (losses) to the planned economy. As we have seen, the advent of industrial hog production created a new social relationship between these hogs and the people who managed them. This was true not only for pig keepers on the farm but also for planners like Felfe, who tracked industrial hog health and maturation from the heights of the planned economy. Felfe never saw individual hogs on his desk; when he managed them, he did as a single herd or *Bestand*—the aggregation of approximately twelve million porcine lives.[23]

The planner's view, however, was ambiguous, especially in the case of the mass die off. That ambiguity was expressed through the use of the term *Verluste*, which farmers and planners alike used to describe production shortfalls. It became an inexact concept when it came to measuring the actual severity of the pork crisis. Planners generally used the term to describe the difference between expected production and actual production. For example, if the planned number of finisher hogs, or *Mastschweine*, for the first quarter was one million and planners actually counted nine hundred thousand, that meant a "loss" (*Verluste*) of one hundred thousand, or 10 percent. Those losses, however, did not necessarily mean that one hundred thousand pigs died, but rather that one hundred thousand pigs were missing from the expected weight class. The losses could have resulted from pigs that died in lower weight classes or pigs that were very much still alive but had yet to put on weight. When planners wanted to denote animals that had actually died, they used the term *Verendet*, or "perished." Yet most of the time, they preferred to elide the difference and to refer to the umbrella term of losses.

When not counting pig numbers, they also used "losses" to describe changes in meat production. Discussing live animals and meat at the same time without distinguishing between them made the amorphous term *losses* even vaguer. The verbal sleight of hand made it more difficult for planners to determine whether losses were due to disease or a fodder shortage—but it made it easier for farmers and planners to cover up real shortfalls. For example, if a farmer reduced his total stock yet raised on average fatter animals, the total amount of meat produced might actually remain unchanged or even increase. Losses in meat production could also be made up through substitutions from other types of meat. Planners might declare a production quota met, even with a significant drop in the amount of pork produced. For example, if a district (*Kreise*) council discovered that its farms were short ten kilotons of pork, it could reach its target by finding an additional ten kilotons of chicken to make up the difference. From the planning perspective, the production quota appeared fulfilled. As we will see, the inexact concept of losses allowed planners to obscure the depth of the pork shortage and thus its ultimate causes.

By early February 1982, however, the dimensions of the crisis had yet to work their way up the economic planning ladder. Planners believed that they had identified the cause. For Felfe, the structure of the losses—in particular the losses in piglets and *Mastschweine*—offered major clues. The decline in numbers of the fattest/most mature hogs pointed to increased demand for pork, while a similar dip in the youngest pigs—piglets and weaners—suggests fodder shortages and sow malnutrition.[24] In other words, the fingerprints of Mittag's pork trade deal and Schürer's fodder import cuts were all over the crisis.

Fodder shortages did not affect all pigs the same, as lifecycle feeding tailored pig diets by their weight and age class. The fattest pigs, the *Mastschweine*, grazed mostly on high-caloric fodder, while piglets, weaners, and feeder hogs (*Läufer*) required more specialized nutrition. The latter two groups had been weaned but were still developing mature digestive tracts and bodies. Weaners and feeder hogs require more vitamins, minerals, and nutrients, which they got through their protein-laden rations. Without the scientifically calculated diet, a fodder shortage would, with luck, only slow growth rates; it could also kill stock at critical development points. Under the dual pressure of fodder shortages and production quotas, many collective farmers slaughtered younger animals. According to one veterinarian, "These losses are partially due to the increased slaughtering of under-weight pigs, and also due to the insufficient growth in the stock of 50–80 kg weight class."[25]

Felfe and the Ministry of Agriculture traced this pattern back to the previous year. "The decline in the population of grower pigs of the highest weight classes has resulted from the overproduction of the plan of 1981. Since the fall of 1981, the poor fodder situation, especially in the month of December, slowed the daily growth rates of hogs, and it led to increased interventions in the stocks, as well as the slaughtering of grower hogs with lower average weights."[26] The farmers' decision to slaughter underweight pigs, likely in fulfillment of Mittag's pork deal, essentially borrowed January's grower hogs in order to make December's quotas.

Under ideal industrial circumstances, East Germany's collective farms might have weathered the disruptions but they occurred on top of already poor environmental and agricultural conditions—the result of the incomplete development of industrial facilities after the implementation of the Grüneberg Plan. As one veterinarian reported:

> The problems of insufficient fodder supplies and the connected issue of low (disease) resistance among the livestock have exacerbated the existing mistakes and shortages in the keeping and care of the animals, and have contributed considerably to the rise in livestock losses. The animals are

not in any condition to deal with mistakes like improper fodder substitutions; unsuitable water troughs for juvenile pigs; frozen, moldy, or spoiled fodder; poor barn hygiene and unregulated temperatures inside facilities; as well as little or inadequate animal supervision. They [the animals] fall ill and die quickly. . . . Veterinary care is hopeless when the general conditions are so poor.[27]

A worker and farmer inspection brigade uncovered a backlog of pig carcasses on many farms. This violation of health codes clearly threatened the surviving animals. In the region of Rostock, the report estimated that eighty tons of dead animals, the equivalent of between eight hundred and a thousand pigs were found on farms. In Halle, the number was even higher, with more than 370 tons of dead livestock waiting to be picked up by the state's livestock processing facilities. The backlog forced one farm to bury 120 tons of dead pigs rather than wait for the state.[28]

Felfe and his colleagues stressed the severity of the crisis to the rest of Politburo. They warned that domestic supplies were imperiled. These warnings, however, were ignored by the pro-export camp surrounding Honecker. And so, despite heavy production losses, the SPK refused to let up on exports. In January and in the first half of February, planners managed to overfill the meat export quotas.[29] Despite the downturn in the hog population, the regime remained yoked to its debt obligations. Yet had industrial pork production continued more or less on its alarming but not yet disastrous path, the regime could have met its export obligations while keeping domestic supplies at minimally adequate levels. Unfortunately, an epizootic outbreak in March assured the disaster would only get worse.

MARCH: THE FOOT AND MOUTH OUTBREAK

While Werner Felfe and his staff at the Ministry of Agriculture were assessing the fodder shortage and analyzing statistics of livestock losses, the chairman of the LPG T Murchin was making his daily rounds of his facility's barns, 175 kilometers to the north in the region of New Brandenburg. The LPG T's livestock units were spread out over a twenty-kilometer area that transected the district of Anklam, so the Murchin chairman drove from barn to barn, perched on ridges above the Anklamer valley. Farms, moors, and marshland all drained to the Peene estuary to the northeast. Across the narrow waterway lay the amoeba-shaped island of Usedom, the second largest of the East

German islands after Rügen, and a land formation that the GDR shared with Poland. To the north lay the icy early spring waters of the Baltic Sea.

In March, the chairman noticed a change in his livestock. A couple of steers were suddenly lame. Others in the Murchin unit had lost their appetite, and some cows were giving less milk. These symptoms by themselves were not that unusual, especially in the wake of the fodder shortage. What was unusual, however, was that a few of the same cattle were refusing to move, while others had begun drooling. And then finally came the telltale sign, when painful blisters erupted on the feet, udders, and mouths of the symptomatic animals.[30] The chairman of the LPG Murchin called his district superiors. On Saturday, March 13, a group from the Regional Veterinary Office of New Brandenburg arrived at Murchin and tested the livestock for the devastating picornavirus that causes foot and mouth disease (FMD).[31]

That day, the veterinary authorities officially confirmed an outbreak of FMD Type O in two facilities of the LPG T Murchin, the first appearance of the disease in two years. By evening of the next day, thirteen young cattle had been diagnosed with FMD out of 399 in the facility. The livestock farm kept 4,344 cattle spread across the villages of Murchin, Runnow, Johannishof, Libnow, Lenschow, and Relzow. The veterinarian investigating the outbreak noted that identifying the disease was only the first step. Foot and mouth disease already had at least a ten-day head start, as the incubation period allowed the virus to spread undetected not only between the small villages of the LPG, but also beyond the borders of the district of Anklam. For example, four days before the veterinary team arrived, 112 heifers had been delivered to six different LPGs across the district including a two-thousand-cow dairy.[32] From there, FMD could spread through milk, but also on the fodder delivered between grain and livestock farms, as well as on tire treads, agricultural equipment, boots, or clothing. At the same time, calves and young cattle from suspected LPGs in Anklam had already been collected and sold. Hence a severe epizootic threat had arisen across the entire district.[33]

An FMD outbreak would be bad enough, but planners had to first rule out other possible diseases. For example, in that same week, a case of Aujeszky's disease, or ADV, had been found in Brandenburg, and wild boars fell under suspicion. Aujeszky's disease had become endemic in a small portion of the country's livestock farms since an outbreak in 1976. On average, around twenty to twenty-five pigs became infected every year. Although the level of ADV infection seemed to remain stable year on year, an outbreak was still traumatic for livestock keepers. Entire litters of piglets had to be killed, many juveniles died, and survivors were often lamed or stunted. According to East German officials, an ADV outbreak could lower production levels by a third. Moreover, ADV could flare up again, even after the initial outbreak seemed to have subsided, carried by other farm animals including dogs, cats, and rats.[34] Planners always suspected that wild boars were a transmission path for ADV. Now with the FMD outbreak, the regime couldn't risk a separate epizootic epidemic and ordered "the complete liquidation of the disease in the entire territory of the German Democratic Republic" through the culling of all wild boars in the affected areas.[35]

On Wednesday, March 17, Heinz Kuhrig, the minister of agriculture, briefed his counterpart in the SED, Bruno Lietz, on the threat posed by FMD and the measures already taken. Less than four days after the outbreak was discovered, Kuhrig ordered a complete quarantine of the infected villages and vaccinations for all herds in the region. "The protective measures also include that no livestock be delivered within the district (of Anklam) or out of the area from the most eastern districts of the region of Rostock, to the northern districts of New Brandenburg, to anywhere else." Most significantly, Kuhrig ordered "the provisioning of Berlin with livestock and meat be carried out from other regions."[36]

The next day, March 18, Kuhrig wrote again to Lietz updating him on changes in the last day. Symptoms of FMD had now been formally established in districts beyond Anklam, including Wolgast, Grimmen, and the island of Rügen.[37] Veterinarians now suspected that the infection had spread from cattle to pigs. The same day, farmers in the LPG New Brandenburg reported that their feeder hogs appeared symptomatic. The cross-species infection now endangered the entire livestock industry. Kuhrig's response was draconian. He ordered the liquidation of all infected herds in the territory and prohibited all meat deliveries from the district. He also mobilized 1,500 reservist soldiers from the National Peoples' Army to help with disinfection measures across the suspected zone of infection. The ministry deployed the soldiers throughout every branch and level of the region's livestock industry, including every "dairy, slaughterhouse, insemination station, livestock rendering facility, cold storage, and other livestock handling facilities."[38]

The FMD outbreak only added to the growing pig crisis. Before March, the decline in the East German pig herd was significant, but it could still be managed by the regime. Foot and mouth disease, however, directly endangered the country's trade obligations, and by turn, the solvency of the East

German government. Like all countries in Europe that wanted to trade meat and livestock, the GDR had to register the outbreak with the World Organization for Animal Health in Paris. And when that happened, the country's trade partners would likely cancel their own trade deals. Military deployments, large scale quarantines—these were the tools of disaster response. The FMD outbreak would require a massive restructuring of the domestic food supply and international trade deals. Suddenly, members of the SPK could no longer shrug off the plaintive demands of the agricultural officials. They would have to deal with one another.

Foot and mouth disease was and still remains one of the most feared livestock diseases of the modern era. Beginning in the nineteenth century, state officials developed a set of scorched-earth measures—namely quarantine and culling—to combat the virus, a standard operating procedure passed down to countries around the world and deployed by the East Germans in March of 1982. Foot and mouth disease was not always feared in this way. In the early nineteenth century, British farmers viewed the disease as a relatively mild, albeit disruptive, infection. While some animals died as a result, most recovered in time. It was the rise of bureaucratic states and international markets for meat and livestock that made the extreme measures of quarantine and slaughter the international standard.[39]

In the GDR, the institute that managed FMD research had its origins in this same nineteenth-century period. In 1897, the German government asked the bacteriologist Friedrich Loeffler to begin research into a vaccine for FMD. Loeffler was a part of an international cadre of veterinarians and scientists who had helped codify modern germ theory. A former colleague of Robert Koch, who had raced Louis Pasteur to identify the anthrax bacillus in 1880, Loeffler was famous for his co-discovery in 1883 with Edwin Klebs of the pathogen responsible for diphtheria (named Klebs-Loeffler bacterium). The hunt for the infectious germ responsible for FMD, however, proved more difficult. As Loeffler eventually discovered, FMD did not come from a bacterium, but rather a virus, a member of a vast order of biological agents that were far too small for contemporary instruments to visualize. This made the production of a vaccine exceedingly difficult. In 1910, Loeffler and the German government established a research laboratory for FMD on a tiny spit of land jutting into the Baltic Sea known as Riems. The Friedrich-Loeffler Institute (FLI) was the first research lab in Europe devoted to the disease, and the location where Loeffler's successors developed a vaccine. The success of the

FLI inspired other governments in Europe to establish vaccine-producing laboratories in their countries as well.[40] This was the legacy passed down to the GDR.

At the beginning of April, the Soviet Union dispatched a team of veterinarians and bacteriologists to the GDR to study the outbreak. The team from the FMD research center in Vladimir, USSR, went to the FLI and found the facility in decline.[41] It had suffered serious neglect over the previous three decades. One of the institute staff recalled proudly that he and his colleagues has successfully jury-rigged the automobile engine of an East German Wartburg—the notoriously shoddy competitor to the equally shoddy Trabant—to spin a lab centrifuge.[42] Former staffers admitted after the collapse of the regime that the FLI had produced sterile vaccinations for much of the 1980s, even as the government claimed the country was FMD free.[43]

In April, the Soviet scientists issued a report. "In our opinion the most probable source of the infection is the Friedrich-Loeffler-Institute on the Island of Riems, since all of the first cases of the FMD outbreak occurred along the Baltic Sea shoreline, in close proximity to the quarantine zone." After visiting the FLI on Riems, the Soviet scientists determined that researchers there "had not guaranteed a hermetic seal of the epizootic building in their work with FMD virus, and had failed to install a sterilizing filtration system for the air in the building." They suspected that either the seagulls that fed on the facility's fodder supplies were carriers, or perhaps the "processed" protein feeds of deceased livestock purchased in Riems were responsible.[44] A classified Stasi report from that year pointed to a more systemic and likely source—the lab's wastewater disinfection system, since nearly all of the infected herds lived in close proximity to the Baltic bays and along the coast.[45]

Since the FMD outbreak would catch the attention of international regulators, the regime's anti-FMD efforts went largely toward combating the contagion of fear among trade partners. As news of the FMD outbreak brought trade deals with Comecon countries to a halt, the government worked to delay embargoes by the Federal Republic and Denmark. On March 22, the Ministry of Agriculture informed its trade partners of the drastic steps it had taken to stem the crisis. "In response to our international responsibility to release information about the emergence of foot and mouth disease in the northeastern districts of the GDR, our Hungarian counterparts have halted all East German deliveries of livestock to their country, as well as livestock that moves through their country to the Socialist Republic of Romania."[46] Agricultural planners implored their counterparts in Hungary to ease the restrictions, even asking the Hungarian minister of agriculture to intervene on their behalf.[47]

The FMD crisis created enormous difficulties for East German exports and the country's international partners. It also set in motion a series of state responses, beginning with the enormous quarantine zone, that crippled the planned economy. The district of Pasewalk anchored the southern edge of the quarantine zone, which stretched east from the adjacent districts toward Poland. Heading northwest from Pasewalk, the zone curved up the Baltic Coast toward Greifswald and then turned westward toward the city of Ribnitz-Damgarten. The total area covered around half of the regions of New Brandenburg and Rostock and extended to the large Baltic island of Rügen. It restricted the movement of 350,000 livestock and over 100,000 people.[48]

The quarantine created hardships, not just in the northern part of the country but across the southern regions and in Berlin. It sent an already fragile agricultural sector into a tailspin, as planners had to reorganize the entire meat supply chain to exclude a fifth of the country. Heinz Kuhrig's March 17 order to provision Berlin with meat from other parts of the country touched off a cascade of shortages that rippled through large cities from Rostock in the north to Dresden and Karl-Marx-Stadt in the south.

The FMD outbreak marked a turning point in the crisis year. Before the outbreak, the underperforming livestock farms and falling pig population had been merely a drag on domestic supplies of meat. The downturn was severe, yet not entirely without precedent. Underproduction of pork had occurred before. In the late spring of 1975, a rolling shortage of meat hit Berlin's retail grocery stores and supermarkets, and the capital city's residents complained bitterly.[49] Foot and mouth disease also was not without precedent in the GDR. In 1976 for example, the disease had sprung up in the New Brandenburg but failed to spread beyond two isolated herds. A much smaller quarantine had stamped out the disease by the middle of May.[50]

In 1982, however, the FMD outbreak arose amid a perfect storm of increasing exports and diminishing returns from the country's livestock facilities. It brought the issue of external financial solvency into direct conflict with domestic security. With the unrest of Solidarity fresh on their minds, the East German regime found itself face to face with an impossible choice: should it rescue the country's collective farms and domestic food supply by eliminating exports? Or should it assuage its creditors by carrying on with existing trade agreements, filling the country's coffers with prized deutsch marks and essentially hoping the domestic situation would not deteriorate further?

LATE MARCH TO APRIL: THE IRON LAW OF EXPORTS

On April 28, Willi Stoph, the second longest serving member of the SED, gave a speech to the chairmen of the GDR's fourteen regional councils. Stoph was notorious inside and outside of the country. SED members called him the "party soldier" for his fidelity to the SED line. Others described him as aloof, always "on duty," and incredibly frugal. "He was modest," remembered Schürer. He "smoked cheap cigarettes, wore his suits for years on end, barely drank more than a beer, and saved the coffee and leftovers at the end of Politburo meetings to take home with himself—much to the embarrassment of other Politburo members."[51] By the 1980s, this miserly, abstemious member of the Politburo and chairman of the Council of Ministers rose through the ranks to reach Erich Honecker's inner circle. Despite the crisis roiling the country that April, Stoph threw his influence behind the pro-export camp. There would be no ignoring the so-called iron law of exports, he declared.[52] East Germans would just have to tighten their belts a bit more. It wasn't that hard, Stoph must have thought. He had been doing it all his life.

Since the announced quarantine a month earlier, the top ranks of the regime had been locked in bitter debate. The FMD outbreak had brought the export question front and center once again. Felfe and his gang in the Ministry of Agriculture pushed for cutting exports and restoring grain imports. The State Planning Commission would have none of it. On March 25, Gerhard Schürer, the head of the SPK, delivered a speech at the weekly meeting of the Council of Ministers reaffirming East Germany's commitment to continuing exports, despite the difficult domestic situation. The pressure of finance capital was especially acute at that moment. A week earlier, Schürer's superior, Günter Mittag, had met with the West German economic minister Graf Lambsdorff in East Berlin.[53] The threat of default had become especially acute by early 1982, as the GDR's short-term loans (essentially rolled-over interest payments) made up 55 percent of its total DM 25 billion debt to the Federal Republic. The debt crisis had both short- and long-term causes. The previous autumn, both Romania ($10.2 billion debt) and Poland ($20 billion) had defaulted on their Western loans. When the Polish regime had raised domestic food prices in 1980 to improve its balance of payments, it set off the Solidarity uprising. These defaults by East Bloc partners, coupled with the liquidity crunch in Western markets (the product of Federal Reserve Chairman Paul Volcker's plan to combat stagflation in the United States) had increased the cost of borrowing substantially. As a result, foreign investors called in 40 percent of their short-term loans to the GDR in the first six months of 1982.[54]

At that March 23 meeting, Schürer blamed the pork crisis on the Western countries and capitalism. "The import of fodder has been affected by embargo politics. . . . Our pig stocks continue to diminish because of fodder shortages, but the rapid speed with which they have diminished was not foreseeable."[55] Missing from his analysis was Schürer's own role in cutting grain imports. Schürer and his allies refused to abandon Western trade. Exports had a miraculous hold over Schürer's faction. They were a panacea for the planned economy. "In connection with the situation regarding the exports to the non-socialist economies, the Politburo has decided to continue exports to these countries in order to relieve the domestic provisioning of the GDR. It is essential, in order to acquire the . . . mark, that the GDR resist the pressure of the banks of finance capital. The GDR must never find itself in the position of being unable to repay its bills."[56] Twisting himself in rhetorical knots, Schürer claimed that to acquire the West German currency vital to the GDR's broader trade interests, East Germany had to meet its debt obligations with the "banks of finance capital."

In response to Schürer's March 23 report, the minister of agriculture, Heinz Kuhrig, pushed back. Kuhrig laid the crisis directly at the feet of the SPK. "The high commitment to over-produce the 1981 plan by 194 kilotons of livestock and 334 million eggs have led to the shortage."[57] The Ministry of Agriculture (MLFN) saw the dependency on volatile foreign markets as a direct cause of economic instability. For example, in late 1981, a fall in international pork prices forced the country to export more pork to make up the monetary shortfall. Kuhrig pointed out that fodder funds reserved for the entire economic year 1981–82 had been almost entirely dispersed in the second half of 1981, at the expense of the first half of 1982. This amounted to a loss for 1982 of around one million tons of grain. The problem of increasing pork production was exacerbated further by a poor grain harvest, which came on top of the planned reduction of grain fodder imports from the West. The result, said Kuhrig, was disastrous. "In the first half of 1982 there was 5.2 million tons of concentrate fodder (grain, mixed fodder, potatoes, and beets) available. That is 1.5 million tons, or 22.4 percent less than in the first half of 1981."[58] The MLFN seemed to argue that without these onerous credit obligations, East Germany's livestock farmers would not have been forced to produce more meat, thereby using more fodder and opening up major gaps in the country's meat and grain supplies. For the MLFN, the link between the collapse of the hog population, the fodder shortage, and trade obligations was clear.[59]

Predicting severe domestic shortfalls of meat and butter, the MLFN made several recommendations. Poultry substitutes could cover thirty-nine kilotons of the meat shortfall while stopping pork exports would add another forty-one kilotons toward filling the domestic supply gap between April 1 and December 31 of 1982. The MLFN also argued that while increasing grain imports was not ideal, the consequences of not importing more grain were even worse. Without further grain imports, the MLFN predicted that under current conditions, the meat shortage would be grow to 17–23 percent of the planned meat production for the nine months between April 1 and the end of the year. The report concluded, "Based on these serious consequences for the provisioning of the population, it is recommended that no further reductions to fodder imports or increased exports of food take place."[60]

With backing from the GDR's chief economic minister, Günter Mittag, and the miserly Stoph, Schürer urged his colleagues to reject the MLFN's proposal. "The measures contained in their proposal call for further, 'above' the plan imports from the non-socialist economies [NSW], for which there is no financing, and the simultaneous suspension of all exports to the NSW, including the Politburo's March 10 resolution for additional exports of meat and butter. The cut in exports alone would represent a loss of 400 million *Valuta* marks hard cash." If not for their current trade agreements, Schürer argued, things would be much worse. "In the People's Economic Plan of 1982, the intended imports of grain and fodder unfolded as planned in the first quarter, despite the major difficulties in financing. . . . At no time did we cut the imports (of fodder), but rather it was a problem of financing, whose fault alone lies with the credit markets. This problem, however, will only get worse now that the MLFN wants to block meat and butter exports. According to the assessment of the SPK this is no solution to the problem, but rather a further intensification."[61]

By the time Party Soldier Stoph addressed the Council of Ministers a month later, the domestic situation had not improved. Yet Stoph argued that the regime should dip even further into domestic supplies to honor its commitments to foreign trade partners with the promised pork deliveries.[62] Ironically, by the time of Stoph's speech, trade quotas to the Soviet Union had not only been met by early March 1982 but overfilled by 129.2 percent. At the same time, agricultural commodity exports to the NSW remained stuck at 78.3 percent.[63] Despite the production difficulties, declining livestock numbers, and the backlog in meat exports to the West, the overall volume of traded goods was still greater than the previous year's, reaching 111 percent of 1981's overall output. Exports to the NSW, while off by 11 percent of 1982's target, still exceeded 1981's numbers by 129 percent.[64]

Stoph's speech confirmed what most East Germans had long suspected: not all economic goals were made equal. Instead, there was a hierarchy of needs that privileged export trade above all else. As Stoph explained it, "We must maintain this ranking and order. The delivery obligations for the capital city of Berlin, the provisioning of the industrial centers in other regions, and assurance of exports as policy, are not new, but rather have existed for more than 25 years. Exports go before self-provisioning [*Selbstversorgen*] . . . sometimes a few comrades forget this fact." Stoph emphasized the point especially for the regional chairmen. In a slightly menacing tone, he argued that the current agricultural crisis had not been caused by a failure of basic policies or by the failure of collective farmers to meet trade requirements. No, said Stoph, it was a problem of "political clarity" that could be made more transparent by the sending of "qualified delegates" to regional councils and farms, in order to "elaborate convincingly" the proper measures.[65] In short, make up for the trade deficits or you will be replaced. That April, the pro-export camp of Stoph, Schürer, and Mittag solidified its control over the regime's response to the pork crisis. Even agricultural officials came to embrace this position. In an April 30 meeting, MLFN member Ostmann urged his boss Felfe to "struggle energetically to follow the resolutions of the Politburo."[66]

Despite the rolling shortages across the country, the petitions and complaints from citizens, and the state of the pig population, nothing could break the regime's faith in exports. The fact that Honecker, Mittag, Schürer, and Stoph were all willing to overlook a growing domestic economic crisis—one that endangered a central platform (cheap, plentiful food) of Honecker's main social policies—shows how tightly woven the East German economy had become in world trade markets. April and May would put their faith to the test, as East Germans pushed back against the regime's imposed austerity.

APRIL: NO PORK CHOPS FOR EASTER, NO WURST ON MAY DAY

April 1982 was exceptionally difficult for restaurants, grocery stores, and retail shops. The Politburo's March 30 resolution left many cities short of pork. Since the Schürer/Mittag camp refused to address the domestic shortages by curtailing foreign trade, the MLFN and the associated ministries of food production were left to cope on their own. Their first move was to conserve existing supplies. They did this by altering formulas and recipes for food manufacturing. They also changed delivery schedules, responding to moments of peak demand or shifting supplies to areas most often afflicted by shortages.

The FMD quarantine compounded the already complicated logistics. Distance, storage, and processing capacity were always critical factors even in normal years. The FMD quarantine threatened to cut off the north, where farmers raised most of the country's livestock, from the center and south, where the vast majority of the country's slaughterhouses were located. The plan had always been based on a smooth-running, fixed infrastructure of rail plus the perfect coordination of livestock delivery, slaughter, storage, and distribution. Livestock was usually carried from the farm, either directly to a local slaughterhouse or to the *Reichsbahn* (railroad) stockyards. Farmers undertook between 20 and 30 percent of all livestock deliveries, while slaughterhouse workers carried out the rest. Some livestock were shipped by train to slaughterhouses within their own region, traveling no farther than fifty kilometers. By 1982, most livestock followed a north-south movement, shipped from the less-populated north to Berlin and southern industrial centers. Cattle from Rostock and Potsdam were sent to the slaughterhouse at the Eberswalde complex, while pigs went to facilities in Berlin and Karl-Marx-Stadt. Pig farmers in New Brandenburg sent their hogs to Erfurt and Berlin's slaughterhouses.

The flow of livestock to market was often slowed by bottlenecks and logjams at the slaughterhouses. Of the seventy-six southern and central slaughterhouses, fifty-nine (78 percent) dated from the nineteenth century. Only six had been built since 1954. The only modern facility was Eberswalde/Britz. Even if the slaughterhouses were running on time, there was no guarantee that the meat they produced could be safely stored. State inspectors found that the Grüneberg Plan had caused production to rise faster than storage capacity. As of 1982, the VVB KLW (state-run refrigerated warehouses) could "only handle 90 percent of the GDR's livestock."[67]

The FMD quarantine threw this system into chaos. On March 24, Felfe ordered the general director of the VVB KLW to reorganize the entire meat production chain. The first order of business was to shift supply sources so that Berlin would not go without. Yet resupplying Berlin proved difficult. With the meat supply already critically low, the entire system of provisioning became a zero-sum game: one city's gain was another's loss. When

Berlin's supply from the slaughterhouse in Pasewalk near Anklam was cut off, the director of the VVB KLW ordered slaughterhouses in the northwestern region of Schwerin to supply an extra delivery for the capital city. This "balancing-out decision" (*Bilanzentscheidung*) was countered by a similar move to replace the lost meat for the city of Rostock, which had previously been supplied by farms in Schwerin.[68] Now Rostock's loss would be made up with a delivery of pork, previously designated for Berlin, from the neighboring Baltic city of Wismar.[69]

Faced with shortfalls, planners from the MLFN and SPK proposed alternative pork substitutes, offering poultry and beef. Gerhard Schürer said the whole shortage crisis was overblown. "All together there are 10,000 tons less than previously known. This [shortage] is clustered in and around a few regions. Yet when one considers the total demand for meat in the GDR, this number seems insignificant." Therefore, he argued that few would notice if restaurants altered their menus to reflect the seasonal shortages. "Hardly any other country has an annual demand for meat of 90 kg per person, like the GDR. Based on this high ranking we must run a smart provisioning program. It is normal to offer five separate dishes on a restaurant menu, where one of the dishes is made without meat at all and another is made with poultry. We must introduce such measures into our own establishments."[70]

East Germans, however, would have none of it. The peak months of shortage—April and May—were also the peak season of pork consumption. Spring holidays and special events cried out for pork with Easter on April 11 that year, followed by May Day, Pentecost (May 30), and youth graduations and confirmations (the secular alternative in the GDR was known as *Jugendweihe*). To celebrate, East Germans demanded traditional foods, and there was nothing more traditional than pork. Substitute meats wouldn't do on such big days. If pork was available or rumored to be available, people queued up. The regime failed to account for stubborn fidelity of its people to their traditions, especially to spring pork. The result was increasing shortages and long, long lines at food stores.

In Berlin, the situation varied by neighborhood. Regional planners reported a placid situation to their superiors, albeit with rolling shortages throughout the city. The directors of Berlin's warehouses had a bird's-eye daily view of the supply situation. Between April 1 and April 20, Berlin planners inspected fifty-five supermarkets and fifty-seven shops of "goods of daily need" (*Waren der tägliche Bedarf*). In general, the directors concluded that "the provisioning of meat and sausage products can, in most cases, continue to be carried out at the existing level." Yet in the very same report they noted widespread problems. "In about a third of the supermarkets and shops, supply gaps for a few different types of goods arise regularly, which can be tied back to a few cases of 'subjective misconduct.'"[71] The "subjective misconduct" seems to refer to hoarding practices East Germans had longed used to circumvent shortages. A prime example was so-called *Bück dich Waren* (bend-over goods), where rare goods were set aside in stores or factories and reserved for family members or close friends. A family member who worked at a butcher's shop might set a quality cut of meat under the counter, bending over to retrieve it when the family member came to the butcher shop. Personal connections ("vitamin-B," for *Beziehung*, or relationships, as East Germans referred to it) played an enormous role in the planned economy for decades and condemning them in this 1982 inspection report was nothing out of the ordinary.[72] In fact, official moaning about "subjective misconduct" masked the severity of the broader structural problems besetting the planned economy that April.

Warehouse directors were well aware of the dynamic relationship between Berlin's consumers and their supermarkets and retail stores. They distributed groceries based on population density. This carefully planned distribution, however, came apart when Berliners themselves figured out which stores were given priority. Because of Berlin's special status as the capital city and the showcase to the West, East Germans knew that the city center was usually well provisioned. Whenever shortages emerged outside Berlin, as they did in 1982 in places like Potsdam, the Potsdammers headed for the capital, stopping at the first neighborhood shop inside the city to stock up. As a Berlin regional planner explained to his boss, "There are several problems in the region of Frankfurt am Oder. In a few districts, like Bernau, Straußberg, Fürstenwald, and Erkner, grilling meats and pot roasts have dwindled in supply on Thursdays and Fridays. The supplies last for two or three hours and are made available for sale mostly in the evening. There are usually four to five varieties of reasonably priced cuts of meat and ten to fifteen types of sausage, most of which are Leberwurst."[73] When these Berlin stores received quality products, the supplies lasted for no more than a few hours, and so the impression of shortages grew. This set off panic buying of pork in Berlin's neighborhood stores, which in turn sent savvy Berliners into the city center to queue for pork products, in front of gawking Westerners. This embarrassed the regime, forcing the state to resupply these visible outlets ahead of schedule, which in turn created more shortages in other parts of the city. And on and on the cycle went.[74]

Women were to blame, declared planners. "In all discussions of these questions, attention is given to how prevent the increased hoarding of a few citizens (so-called occasional customers) and how to assure the appropriate distribution of goods to everyone. Repeatedly it is stressed that in the retail stores of rural villages, housewives line up for the meat delivery truck during the day and buy the best cuts immediately, while working women, who have to shop after work, only have the leftovers to pick through."[75] In response, some citizens suggested a return to the rationed provisioning system of the 1960s, when every household was assigned a butcher shop or retail store as the only place they could buy meat.[76] This gendered analysis revealed the increased economic and political power of women, and as a result, the seriousness with which planners took women's complaints.[77]

Similar irregularities in meat supplies also emerged outside of Berlin, especially as the spring holiday season approached. In the regions of Dresden, Karl-Marx-Stadt, and Rostock, planners knew that peak shortages correlated with the Easter Holiday, weekends, May Day, and Pentecost. Now planners worried that the vacation season would exacerbate the situation further, when many people left their urban housing blocks for allotment gardens and country homes for outdoor grilling parties. This drove peak demand, especially as the weekend neared.[78] At the end of April, Heinz-Ralf S., a resident of Leipzig, petitioned the regime, describing how the meat shortages were affecting his daily routine. "Recently meat products and sausages have become not only scarce, but also only partially available. The butcher shops are empty. Stores are receiving across the board cuts in supply. It looks especially bad in grocery stores, as employees have become discouraged. Today at work many people were complaining about the state of our lunch menu and as well as breakfast, which (we heard) will be shortly discontinued. The cooks want to stop so they don't have to listen to the workers' complaints anymore. . . . The poor situation with meat is being discussed everywhere."[79]

The regime reduced the problem to fixing delivery schedules. Noting that certain meats were purchased at the beginning of the week for family meals while others were in demand at the end of the week for social gatherings, planners attempted to align their deliveries with consumer patterns.

> Several regional committees, e.g. in Dresden, Cottbus, Halle, Suhl, Leipzig, and Schwerin, have estimated that no continuous supply can be guaranteed every day of the week. Mainly sausages and pot roasts are available during the first half of the week while grilling meats are

> concentrated in the second half. This information, in agreement with the estimates of many regional councils, largely coincides with the actual demand of citizens, but has the unintended effect of creating lines (outside stores) in the second half of the week. Deliberate measures, like concentrating deliveries toward the second half of the week and re-apportioning servings have provided little noticeable relief.[80]

At times, it was hard to determine the true level of demand. In every region, meat shortages seemed to take on their own special character. In Dresden, despite every effort of the regional council to maintain a constant supply, availability was unreliable and unpredictable. In some areas, panic purchases set off runs on meat products. Districts in Dresden reported women leaving their factory jobs early to catch the daily meat deliveries. In Cottbus, meat supplies seemed to be on every mind and every tongue. The same conversations were going on in Halle, where hoarding was reported on the rise. The regional leader of Suhl told his superiors that many citizens were getting their meat-crisis news from the West German media, not from the authorities. This was accelerating hoarding.[81]

While waiting in line for meats, people increasingly turned their ire toward the ongoing exports of pork to the West. In Leipzig, the discrepancy between what authorities were reporting and the truth upset residents even more. The regional chairman reported that "it is expressed with little understanding, that the holes in supply, especially of meat, have led to a decrease in the number of meat dishes in the largest restaurants, even as our mass media always reports how the plan has been overfilled with ever higher results."[82] Others blamed the exports and the outbreak of foot and mouth disease. Some regional chairmen suggested a compromise that would swap non-meat exports for precious pork. "From our perspective," the chairmen of the regional council in Suhl wrote to Bruno Lietz, "I would recommend that instead of the exports of meat, we export extra wood, lumber, and other valuable industrial goods like mopeds, so that the disturbance of domestic supply is not so acutely felt."[83]

MAY, JUNE, JULY: STABILIZATION AND CONTAGION

After the spring holidays, the pork crisis began to stabilize. The situation in the northern regions improved. Spot inspections in the districts of Ludwigslust and Perleberg in the region of Schwerin showed marked improvements in the meat supply in several towns and cities.[84] By the end of May, the overall pig losses appeared either less severe or stable at around a million animals. June looked even better. On June 30, planners reported monthly pig numbers on par with their early year projections. While overall production remained down for the year, collapse had been averted.[85] Monthly losses among piglets and weaners had been cut in half in the first six months of the year, stabilizing the crisis. On August 31, Heinz Kuhrig wrote to Werner Felfe, crediting farmers and agricultural planners for halting the downward spiral. "The unprecedented rise in livestock losses for the first quarter of 1982—especially in piglets, weaners, grower hogs, and calves—was kept within the bounds of 1981 levels by the end of the second quarter through the shared efforts of collective farmers and workers, and with the support of veterinarians."[86]

Spring and summer also helped alleviate the crisis, although none of the planners said so explicitly. The two greatest threats to livestock in the winter—fodder shortages and disease—became less serious as the weather turned. On many farms, livestock moved outside, thus reducing the spread of disease, while many cattle farmers let their animals graze on newly verdant pastures. Kuhrig believed that the agricultural sector overcame these problems only when the district and regional councils began to coordinate with animal caretakers and veterinarians. The MLFN recommended that the councils organize training seminars for agricultural workers, increase the number of veterinarians in the country, and encourage grain farmers to become less specialized so that they could help out with livestock production in the difficult winter months.[87]

The stabilization of the pig population, however, did not bring an end to the shortage crisis. The winter disruptions spilled over into complementary sectors of the planned economy. The foot and mouth quarantine in March forced planners to adjust the production quotas not only for meat, but also goods that required milk, butter, and animal fats. To adjust for the shortages, planners proposed a series of drastic conservation measures. They suggested lowering the percentage of meat in certain goods like schnitzels, sausages, and roulades, increasing the supply of poultry in the stores, and ordering food producers to alter their recipes. Shortfalls in the country's supply of raw fat put sausage makers behind. Kuhrig recommended that factories reduce "the fat percentage to 3 percent in all kinds of wurst, and a change the types of wurst available."[88] In other industries, planners urged brewers to use less malt, bakers reduce the size of their loaves and cakes, and that all food producers economize on their use of sugar and fat.[89] The crisis had also hit cheese and butter factories. To conserve milk fat, the Politburo ordered changes to industrial formulas. No detail for lowering costs was too minor, as they also ordered that less ink be used in the printing on the side of containers.[90] Later in the year, planners changed the formulas for butter production by watering down or skimming milk.[91] Similar adjustments to other foods were ordered repeatedly throughout the year. All failed to alleviate the public perception that the GDR was in the grips of an ongoing crisis.

Agriculture and food production limped along for the last three months of 1982. Butter, sugar, and pork were still scarce. Farmers reported a reoccurrence of FMD on the peninsula of Darß. Fortunately, the facility was already under observation from the outbreak earlier in the year. In the fields, the harvest of 1982 revealed a typical paradox of specialized production: while the grain harvest had reached a record high of more than ten million tons, a grain fodder deficit of around two million tons remained.[92] With the prospect of the 1982 holiday season approaching and the crisis still dragging down agricultural production, Mittag and Schürer relented. In October, Mittag withdrew some butter and eggs intended for export, remanding them for domestic consumption.[93] In November, planners purchased an extra twelve kilotons of meat from its eastern neighbors, even as their own *Mastschweine* reached their planned weights. On December 15, the GDR completed a deal to import goose meat, a holiday favorite, from the Soviet Union.[94]

November also saw two momentous changes in regime hierarchy. On November 9, Heinz Kuhrig resigned from his position as minister of agriculture and was replaced by his counterpart from the SED, Bruno Lietz, who served as the chairmen of the German-Soviet Friendship Society, a far cry from credentials that would suggest a powerful position in the planned economy.[95] Ironically, "German-Soviet Friendship" underwent another kind of change the very next day, as Leonid Brezhnev died in Moscow. Kuhrig's demotion could easily have been construed as punishment for his outspoken criticism of the Western exports. The same week, planners reported a total of 1.7 million pigs "lost" in 1982, of which more than a million were piglets.[96]

In a year of frustration, panic, and disappointment for the Ministry of Agriculture, there was a silver lining. Way back on February 1, with industrial agriculture at a precarious tipping point, a state slaughterhouses in the region of New Brandenburg reported a curious phenomenon among their deliveries: while nearly every age class of industrial pig was struggling to maintain weight, garden pigs raised in the backyards of farm households arrived for slaughter more than thirty kilograms over their target weight.[97] The desperate agricultural planners saw a potential support for sagging production quotas in these garden pigs. Faced with the collapse of industrial pig production, the Politburo debated a major overhaul of agricultural prices at their February 2 meeting. They considered a price reform that would requisition the millions of pigs raised individually around the country.

The regime had already outlined a plan that would encourage small producers to bring their fruits, vegetables, and meats into the state economy. "The material incentives for this kind of production must be directed at individual producers, since their traditions and personal interests are more closely aligned with this type of work, while the labor structure of socialist farms cannot produce as cheaply foods like specific vegetables, berries, honey, rabbits, geese, and turkeys."[98] The new prices would also extend to beef, eggs, and most significantly, pork. Prices the state paid for live animals under the new system would rise to four times the previous amount. The government hoped to equalize this price reform by also increasing the price of grain and protein fodder for the animals.[99] It wasn't until the fall, however, that the plan was formally codified. On October 26, the Politburo passed the price reform, which raised the state purchase prices for individually produced agricultural goods including pork but also fruits and vegetables.[100]

Reading between the lines, the price reforms masked a startling admission: the MLFN was acknowledging that the country's socialist farms were not working. They hoped that the new prices would encourage gardeners, rural residents, and collective farmers not only to increase output but to sell more of their food to the state, rather than consuming it with their friends and families.[101] In other words, the gigantic factory farms were failing in their basic task of providing a steady supply of affordable food. This was a de facto acknowledgment that there were limits to the factory farm, tantamount to a repudiation of over thirty years of agricultural development along industrial lines. The price reforms would go into effect on January 1, 1984.

PIG MEN

In the final section of George Orwell's 1946 allegory *Animal Farm*, the leaders of the animal revolution—the pigs—finally walk on their hind legs, violating their fundamental revolutionary tenet of "Two legs bad, four legs good." As the other barnyard animals stare slack jawed, their swinish leaders lurch bipedally out of the old farmhouse and across the yard. This is the final transformation of the pigs from comrade liberators of the downtrodden animal masses to two-legged imitations of the farm's onetime master, the cruel and odious Mr. Jones. The pigs also wear human clothes, brandish whips, sleep in beds, and drink alcohol. While the pigs strut about, the other farm animals subsist on minimal rations while performing maximum work. Everything the animal revolution once opposed—inequality, cruelty, and drudgery—has returned to Animal Farm.[102]

After the crisis year of 1982, the rulers of East Germany seemed to stand on Orwell's metaphorical two legs. Like Napoleon the Pig, Erich Honecker filled his private life with the finest capitalist goods, including an inexhaustible supply of HB cigarettes from West Germany, a Citroën limousine, and a garage of late model Western SUVs. Honecker's hunting lodge and private game preserve rivaled Stalin's dacha at Kuntsevo. Gunther Mittag walked on an artificial leg made in Japan and purchased a dozen Sony color televisions. By the 1980s, the regime was spending between DM 6 and 8 million every year in the West to fulfill the Politburo's private shopping lists.[103]

The crisis marked a decisive victory for the pro-export camp. For members of the SPK, industrial export agriculture was a preview of a future GDR, proudly independent of Western markets. But if 1982 offered a glimpse of this future, they did not like what they saw. In the near term, there would be no transformation into an "export land." East Germany's farms, pigs, and waters were now enmeshed in global flows of cheap commodities that serviced the expansion of industrial agriculture everywhere. They would absorb the environmental costs of a globe-spanning factory farm for the rest of the decade.

Even if the regime had wished to return to the status ante-1982, they still needed to find new sources of capital. In the summer of 1983, West Germany rescued East Germany. The two countries surprised world markets by announcing a DM 1 billion, interest-free loan. Even more shocking were the figures involved. The previous year, Helmut Kohl and the Christian Democratic Union swept the Schmidt government out of power. As the new chancellor of the FRG, Kohl brought Franz-Josef Strauss into the government. Strauss had risen to fame and political power in the 1960s as an ardent anti-communist but was forced to resign following a political scandal in 1962.[104] Strauss spent his years in the political wilderness building up personal connections with the Eastern Bloc. His triumph was arranging loans for a pipeline deal with the Soviet Union in 1969–70.[105] By 1983, Strauss had made friends with key members of the East German regime, including Alexander Schalck-Golodowski, the shadowy head of the Commercial Coordinating

Ministry, or KoKo. The covert action organization's chief task was to find as many illicit sources of West German hard currency as possible.[106]

It was these two men, Schalck and Strauss, who put the billion deutsch mark credit deal together. Through the first half of 1983, they met in secret to work out the terms. Financed by Bavarian Landesbank with the full backing of the Federal Republic, the deal was announced on June 5.[107] Strauss was roundly criticized in the West German press, even as he continued to insist the deal was made to improve German-German relations. In the short term, the deal had to be bolstered by a second round of capital infusion the following year with the East Germans receiving an additional DM 950 million. The backing of the FRG, however, essentially removed the risk factor and allowed the GDR to seek loans from other foreign banks.[108]

Underlying the loan, however, was an agricultural deal. One of Strauss's closest friends was Joseph März, a Bavarian cattle dealer, who had done business wholesaling meat to the KoKo. Strauss's return to power presented an opportunity to März, who was quickly introduced to Schalck.[109] The three men worked together to push aside a rival meat wholesaler affiliated with the Social Democrats and take over the East-West pork and beef trade. The restoration of East Germany's credit, coupled with a warming in East-West relations, seemed to guarantee GDR agriculture a profitable future.

The Bavarian capitalists and the East German fixer developed a close relationship. According to Robert Dale, Schalck and Strauss came to admire each other. While Strauss and März found the East German's Prussian authoritarianism appealing, especially the regime's suppression of "hashish, pornography, and long hair" among the East German youth, Schalck, Honecker, and Mittag admired their West German counterparts. In his memoirs, Mittag wrote how grief-stricken he became when he heard of Strauss's death in 1988.[110] Honecker's own pull to the West had brought the East German economy into full alignment with the capitalist class.

West Germany's rescue of East Germany's animal farms would have been no surprise to George Orwell. *Animal Farm's* dénouement occurs when Napoleon summons Mr. Pilkington, the owner of neighboring Foxwood farm, to a feast. Pilkington, the erstwhile arch enemy, arrives with his men to celebrate the normalization of relations between the farms. As the men and pigs gather for a banquet in the farmhouse, the hungry and cold farm animals outside peek in through the windows. Pilkington and Napoleon take turns praising each other and declaring their past disagreements at an end. Men and pigs play cards and smoke cigarettes. They toast each other and toast the new name of Animal Farm—now reverting to its former name of Manor Farm. No sooner does peace arise between pigs and men than it vanishes in a drunken brawl. Watching the fight from the dark, the farm animals could not see who was winning. As Orwell writes, "The creatures outside looked from pig to man, and from man to pig, and from pig to man again; but already it was impossible to say which was which."[111]

AFTERWORD

Garbage Dump of the West

AN UNOFFICIAL COLLABORATOR WAS LIKELY THE FIRST TO SPOT THE 30 × 30 centimeter notice nailed to the door of the Church of the Brethren (*Brüderkriche*) in Altenburg.[1] As the Stasi needed to know about it, the local police reported it immediately, and by the end of the day, September 14, 1988, the security service was scrutinizing the hand-scrawled message, which began:

> solidarity needed
> on the 02 October at 1400 during the open-air harvest festival service in plothen (district schleiz) near dittersdorf in the surrounding area of the one-hundred-seven-five-thousands pigs of the gigantic industrial finishing facility.
> this facility does not [only] employ animal cruelty methods but also destroys the entire landscape.
> massive forest death, total poisoning of the air and drinking water, sickening residents especially children, surrounding villages are evacuated, 80 percent of the animals are designated for export to the west. we are doing all of this for foreign currency? more information is in the environmental library.[2]

The notice was the work of an environmental group in the nearby villages of Knau and Dittersdorf. The 1988 Harvest Festival service was to be their first public action against the *Schweinezucht und Mastanlage* (pig breeding and fattening plant, SZMK) Neustadt/Orla, a sister facility to the complex in Eberswalde. Local concern and informal discussions about pollution and the

SZMK had been bubbling up since the 1970s, although it was not until 1985 that strategic organizing emerged within the relative safety of the local Protestant church. The issue: the sudden "forest death" (*Waldsterben*) of eight hundred hectares of woodland surrounding the plant.

Although the Stasi were trained to monitor, and if necessary, to stamp out dissent, the security officers who read the Altenburg notice that day did not realize that a political earthquake was about to begin. The complacency is understandable. The Stasi had been tracking the nascent environmental movement for most of the decade. As late as September 1988, environmentalist groups remained small, diffuse, and uncoordinated. As much as the protester in Altenburg may have wanted, the Stasi did not believe that they had a modern Martin Luther on their hands. They were wrong. Just over a year later, the East German regime and the Stasi were gone.

What the Stasi failed to see was that the Altenburg notice represented an escalation in environmentalist organizing, and thus criticism of the regime.[3] In previous private discussions with the SZMK's director, the Knau/Dittersdorf group had demanded reforms at the factory rather than its closure. They wanted upgrades to manure holding tanks, new technologies to mitigate pollution, and increased access to environmental data.[4] But no longer. The Altenburg notice went beyond the criticism of toxic pollution, maltreatment of animals, and threats to public health, to the third rail of political dissent in the GDR—the economy. To question the role of industrial hog production in trade policy was to question the regime's priorities, and thus its fitness to rule. The Socialist Unity Party's (SED) preferential treatment of Western trade was hardly a secret. Every time there was a shortage, people complained bitterly, writing petitions and grumbling in grocery-store lines about export policies. The upper echelon of the party and State Planning Commission (SPK) pontificated about the "iron law of exports" and "the preservation of the GDR's creditworthiness" to anyone who dared question economic orthodoxy.[5] It was the environmentalist movement, however, that pushed the critique further, drawing a connection not just between food shortages and Western exports, but also between exported meat and agricultural pollution. As the pastor, opposition leader, and founder of the new Social Democratic Party, Markus Meckel, argued in an October 1989 speech, "Nature and the environment have been destroyed on a massive scale due to the irresponsible politics and the poor economy [of the regime]. We have become the garbage dump of the West. Cheap pork is sent to the West—lakes of manure remain here."[6]

Meckel's speech occurred amid the drama of the revolutionary fall of 1989. The East German uprising had begun the previous spring, following Hungary's decision to remove the heavily fortified fence with its neighbor Austria. In response, thousands of East Germans flocked to traditional vacation destinations in Hungary, hoping to cross at once or await permission to enter Austria. The exodus caught the attention of the world, as East Germans gathered at West German embassies in Budapest and Prague. Meanwhile, the communist regimes in Hungary and Poland dissolved; East Germans now wondered if their government was next. At the end of the summer, the Monday Demonstrations began in Leipzig, attracting disparate civil society and protest groups to articulate demands: freedom to emigrate, freedom to remain, freedom of speech, and the end of surveillance. They also wanted a stable currency, better consumer durables and an end to shortages. They demanded clean air, clean water, clean power, and an end to forest death.

Each week, the Monday Demonstrations grew larger and larger. On October 7, the regime celebrated the fortieth anniversary of the founding of the GDR. Two days later, the Monday Demo drew three hundred thousand protesters. The following day, the SED moved to oust Erich Honecker, but by then it was too late. On November 3, the demonstrations spread to Berlin, as more than half a million people protested in Karl Marx Allee and Alexanderplatz. Four days later, the entire government resigned. On November 9, SED member Gunter Schabowski announced on television the end of travel restrictions on television, sending fifty thousand East Germans streaming into West Germany. With that, forty years of communist rule was over.[7]

The sudden collapse of the GDR opened a period full of possibility for all Germans, East and West, that became known as Die Wende, or the transition to reunification. Environmentalism, which was widely credited with bringing down the regime, emerged at the center of German national and international politics during this period.[8] Over the next decade, the reunified government, which became known as the Berlin Republic, was determined to make environmental diplomacy its major international priority. At the 1992 United Nations Earth Summit in Rio de Janeiro, German diplomats took the lead in drafting resolutions on greenhouse gases and carbon emissions.[9] Domestic politics also stressed environmental issues, especially when it came to reforming the post-socialist "states," or *Bundesländer*, of the GDR. The Berlin Republic sponsored major pollution remediation programs for the East, while closing down its dirtiest power plants. It developed wind and solar energy, extended its infamously rigorous recycling programs, and underwrote robust mass transportation infrastructure. By the 2000s, the Berlin Republic was heralding remarkable drops in carbon emissions as the best evidence yet of its leading role in environmental issues. Germany's efforts won for the country and its citizens a reputation as the "greenest nation"—a reputation that only increased following the 2011 announcement of the *Die Energiewende*, or transition to 100 percent renewable energy.[10]

While Germany's environmental advances were real, their origins and causes have become less well understood over time. In particular, reunified Germany's environmental progress since 1990, especially its impressive lowering of carbon emissions, cannot be understood apart from the winding up of East German society and ensuing deindustrialization. While government programs like ecological reconstruction ameliorated environmental damage in some of the worst polluted areas, they did less to address the fundamental structure of Germany's industrial economy. Yet the policies that Western politicians and thinkers criticized the GDR for—its reliance on brown coal and its obsession with "gigantomania" in agriculture—have only increased in reunified Germany over the last quarter century. At the same time, many of the biggest ecological successes in the East, including the establishment of nature parks, reforestation, and wildlife restoration, were built upon decisions and programs begun during the GDR. In order to understand the greenwashing of Die Wende, we must return to the politics of reunification.

In 1990, one of the most urgent tasks facing the new government was how to clean up the lands of the GDR. The most pressing concerns were the reliance on brown coal, the use of nuclear energy, and the environmental damage caused by agricultural pollution. In the 1980s, these issues had preoccupied environmentalists in both Germanys. East and West German groups borrowed language, ideas, and tactics from each other. East Germans read smuggled copies of Rachel Carson's *Silent Spring* and Aldo Leopold's *Sand County Almanac* and created their own environmental texts, including newsletters, such as *Umweltblätter* and *Arche Nova*, that were read by fellow citizens as well as their allies in the West.[11] East Germans set up "environmental libraries" (*Umweltbibliotheke*), reading rooms sheltered in Protestant churches, to disseminate samizdat.[12] While the dangers of nuclear power galvanized environmentalism in the 1970s (and again after the 1986 explosion at Chernobyl), the environmental cause célèbre of the 1980s in both Germanys was the threat of acid rain and its destructive symptom, *Waldsterben*, or forest death.[13] In the FRG, *Waldsterben* was the issue of the decade, fueling the rise of the Green Party and its election to its first seats in the Reichstag. The federal bureaucracy enacted major regulations for power plant emissions in 1983.[14] In the GDR, *Waldsterben* was even more widespread. Ninety percent of the country's forests were affected, including more than five hundred thousand hectares of standing dead trees.[15] The transborder threat of acid rain created opportunities for collaboration.[16] The most famous example was the 1988 guerilla documentary, *Bitteres aus Bitterfeld* (Bitter news from Bitterfeld), which secretly filmed evidence of sulfur dioxide pollution in East Germany's industrial south. The documentary aired on West German television, including in West Berlin, which beamed the regime's worst kept secret to millions of East Germans.[17] When the GDR collapsed, groups in both countries saw their collaboration as instrumental in the end of the regime.

Radical environmentalists believed that reunification offered a chance to remake both Germanys. Many saw the former East, and its dirty energy sector, as a tabula rasa for the full implementation of environmental reconstruction. The think tank Institut für Energetek (IfE) proposed a complete and radical break from a century of thinking on energy production, with a complete conversion of the former GDR to renewables.[18] In 1990, the Institut für Ökologische Wirtschaftsforschung published a comprehensive study of East German environmental pollution compiled by a joint team of scientists from East and West Germany. Throughout the evaluation of the energy sector, chemical production, transportation infrastructure, and agricultural production, the authors repeatedly stressed the potential for a true Wende. "The collapse of the system and the opening of the GDR to Western influence brings a number of chances and possibilities for environmentalism in both the East and the West, but also risks. An opportunity has arisen to correct the failed developments of the past."[19] The authors decried the market system as a false panacea, claiming the West German political system had created more than its share of "environmental dead ends."[20] A sustainable future for the post-socialist East, the experts declared, would not be found in West Germany's past.

Other Germans had different ideas. Leading politicians and media coopted the language of 1989's environmentalists to steer the course of Die Wende. They urged a complete economic dismantling of the GDR. No one did more to articulate this view than Chancellor Helmut Kohl. In television interviews and speeches, Kohl painted a picture of a future East transformed into "*blühende Landschaften*," or blooming landscapes—where people would live and work in prosperity.[21] Kohl declared that the former GDR would have a new birth under "private ownership, competition, free pricing." Its former citizens would blossom under an EU-styled "freedom of movement of labor, capital, goods and services." A new *Wirtschaftswunder* was coming for the GDR, just as it had once come for West Germany in the 1950s and 60s, said Kohl.[22] The chancellor's rhetoric painted the GDR as uniquely polluted. Only nature could heal the desecrated East. For neoliberals eager to privatize the most profitable parts of GDR industry, this rhetoric was a gift. Under the cover of environmental remediation, wealthy West Germans and politically privileged East Germans seized the most valuable assets for themselves and left the pollution, destroyed forests, and work of remaking Germany's industrial economy to everybody else.

The greenwashing of Die Wende was aided by the rapid reunification process. In less than a year, the German governments and four occupying powers agreed to dissolve the GDR, a treaty that became known as the "Two-Plus-Four." The hasty transition created natural momentum for neoliberal economic reform and swept away any chance for deliberate planning or ecological reconstruction. "The rush to unification," as Konrad Jarausch argued, left the process "over reported and under analyzed."[23] Fateful changes affecting currency and privatization were ratified in a matter of months, forcing the speedy dissolution of the GDR. Administrative institutions, laws, factories, and property rights were dissolved, abrogated, downsized, and redistributed.[24] Die Wende set off a massive transfer of wealth from East to West. Within three and half years of reunification, over 92 percent of state firms had been privatized, with 85 percent falling into the hands of West Germans, 9 percent to international investors, and only 10 percent to East Germans.[25] In agriculture, privatization pushed hundreds of thousands off the land and out of farm-related industries. East German farmers and agricultural workers organized mass protests in the summer of 1990 to save their farms. They descended on East German cities, their trucks piled high with unsold potatoes, grain, fruit, and animal carcasses. Without the subsidies of the state or loans from banks, the farmers argued, they could not find buyers for their products. Most farmers soon realized that they had been left holding the bag as the economic system around them collapsed.[26] In 1989, East German agriculture employed 850,000 full-time workers. By 1995, it employed 157,000; four years after that, only 112,000 remained.[27] Overall employment in the former GDR dropped from nine million in 1990 to six million in 1997.[28]

Meanwhile, the rapid deindustrialization of Die Wende bolstered the Berlin Republic's green bona fides to the world. Many of these decisions made urgent sense. In a post-Chernobyl Europe, East Germany's four nuclear power

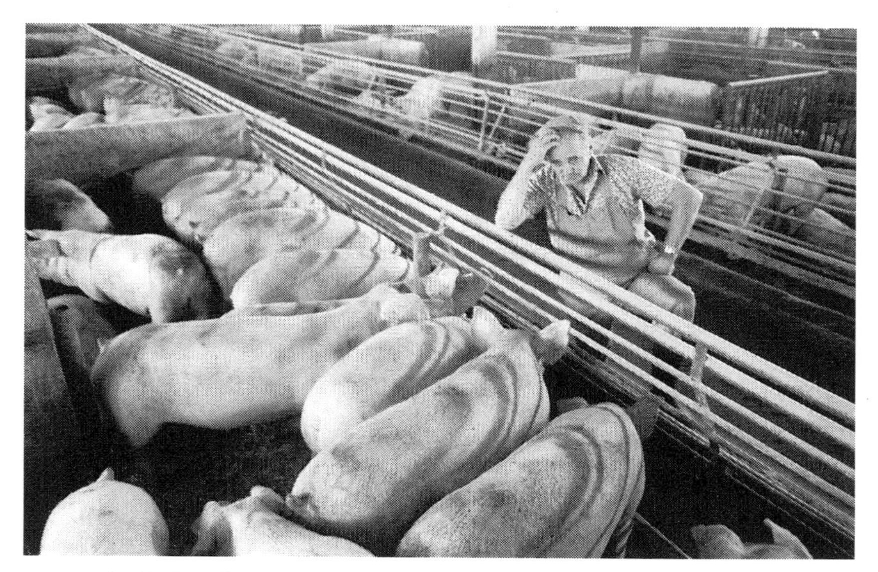

FIG A.1 An LPG worker contemplates his pigs and the future during Die Wende. July 17, 1990. Courtesy of akg-images/ddrbildarchiv.de.

plants were taken offline after an accident in November 1989 at Greifswald, East Germany's largest plant.[29] Brown coal mines and decrepit power plants were targeted shortly thereafter. Had these closures happened under the SED regime, the country would have collapsed. Now the closings went almost unnoticed, as economic activity declined even faster. In this way, environmental progress went hand in hand with economic disintegration. The more factories and mines that closed and collective farms that fell apart, the less energy was consumed. Within two years of reunification, industrial production in the former GDR fell by two thirds, while annual energy consumption declined from twenty-six to sixteen gigawatts. Over the next twenty-five years, carbon emissions in all of Germany fell by 10 percent.[30] The East's deindustrialization gave cover to Germany's energy diplomacy, spreading the belief that economic reforms had been painless and effective. It also allowed Germany to avoid confronting the fundamental structural issues that had caused the GDR's pollution crises in the 1980s and that persisted in the reunified country for the next three decades. This was especially true for East Germany's three pigs.

Today, the areas once occupied by SZMK Eberswalde are half empty. Rows of solar panels stand where the old pavilion-style barns and feed silos once dominated the landscape. Dirt, weeds, and the concrete outlines of the interior roads are all that remain elsewhere. North of the canal, the innovative slaughterhouse built by Berlin-Consult in the 1970s, SVK Eberswalde/Britz, still produces sausage for European markets. In the city center, the famous forestry school that trained generations of East German scientists in 1992 was renamed the School for Sustainable Development. Today, students take courses of study in forest and environment, landscape usage and nature protection, lumber engineering, and sustainable economics.[31] Even in Eberswalde, a mixture of ecological reconstruction and greenwashing has remade the infamous "Hog City."

Yet the GDR's industrial legacy lives on. Rather than abolishing it, the Berlin Republic has doubled down on the grain-oilseed-meat complex by increasing grain imports, maintaining enormous monocultures, and expanding livestock confinement. Between 2000 and 2013, the number of hog rearing facilities decreased from 125,000 to 25,000, while the average number of pigs per facility rose from 350 to 2,163. Over the same period, Germany's annual number of slaughtered hogs rose from forty million to sixty million—production numbers that would have astounded the East Germans.[32]

Germany has also gone from a net importer of meat to a net exporter, finally achieving a long-held dream of German politicians and economists since the days of the Kaiserreich. Yet as in the GDR, the rise in industrial meat production has been accompanied by a dramatic rise in foul, unusable livestock manure. In 2010, Germany produced approximately 191 million cubic meters of animal waste. Some of it leaches out of slurry lagoons and treatment facilities after severe weather. A not insignificant amount—as much as 9.6 million liters in 2015—spills untreated into the environment because of transportation accidents.[33] Livestock facilities, however, are not solely responsible for nitrate pollution. Officials estimate that 37 percent of total nitrogen runoff comes from the over application of agrochemicals. The massive size of eastern Germany's fields, created under Grüneberg, is partly responsible. While the type of crops grown in the former East have changed, the scale of production has not. Nitrate and phosphorous pollution continue to overwhelm Germany's largest coastal areas on the North Sea and Baltic with massive dead zones caused by eutrophication.[34] The industrial pig with its specialized imported diet and its noxious waste is alive and well in postsocialist Germany.

So is the wild boar. In 1991, the German boar population was estimated at four hundred thousand. Today, ecologists say they number around

2.5 million. Encounters between pigs and people are a staple of German newspaper headlines: boars knock old women from bicycles, assault the wheelchair bound, desecrate cemeteries, and to the horror of this dog-loving nation, mortally wound family pets.[35] The boar infestation is not exclusive to Germany. Similar stories can be found in newspapers from Texas to Japan. Across Germany, authorities have organized special hunting squads to cull overpopulated areas. The number of wild boars only continue to grow.

The wild boar problem has distinct roots in the political economy of the GDR, a point recorded in the flesh of around 2 percent of these animals that are contaminated with radiocesium.[36] The fallout from Chernobyl came from the East, the last vestiges of an energy regime produced through Cold War competition. New ideas about conservation came from the West in the 1990s, creating boar habitat all over the country. The 1991 National Solidarity Action for Ecological Restructuring program called for the investment of DM 5 billion for *Sanierung*, or restoration of industrial disaster zones. The Ministry of the Environment quickly identified twelve thousand contamination sites and completed work on six hundred environmental projects while employing fifty thousand people.[37] In the Lausitzer Seeland (the Lausatian lake district), officials transformed fifty-four square miles of what had been East Germany's largest brown coal mining region into a green landscape of lakes and forests. Today, the migratory birds, rare wild flowers, and old growth hardwood forests that ring its lakes belie the park's industrial past.[38] So too do the wild boars that root and roam there.

The latest large-scale conservation project is the European Green Belt. From the Baltic to the Black Sea, this transnational program envisions a contiguous corridor of migration following the footprint of the Iron Curtain, forming a single, continent-wide park.[39] The inspiration for this program came from the surprising discovery in the early 2000s of abundant wildlife and undisturbed and diverse habitats in the demilitarized East-West border.[40] While the program utilizes principles from the community of Western conservation, it is built upon the works of the East German regime. The government of the GDR also set aside several of eastern Germany's current nature parks as protected biospheres (including the infamous hunting estate of Schorfheide) as one of its last acts in 1990—an enduring legacy of state socialism that has aided the proliferation not only of the wild boar but of all kinds of European wildlife.

Not all of East Germany's pigs have fared well. The garden pig has declined dramatically in post-socialist Germany. The shortages produced by the planned economy made the garden pig a necessity. New agricultural regulations were written for industrial farms and animal welfare, but they affected all pigs, regardless of scale or operation type. In 2017, there were no pigs at all inside the Berlin city limits.[41] Yet kitchen gardening and small livestock breeding are growing in popularity for locavores, bio-food consumers (*bio* is the German term for "organic"), and self-reliant city dwellers. Throughout the country, garden clubs and gardening membership have continued to increase. In Berlin, sixty new community gardens were founded between 2004 and 2014.[42] Germans continue to justify their interest in green spaces in traditional terms, wanting to reconnect with nature, produce their own food, or simply relax in their home away from home.[43] More recently, the politics of immigration have spread to German gardens. Proponents argue that gardens serve as a site of cultural integration for Russian, Eastern European, and Turkish immigrants, while racist, fearful opponents claim these spaces serve as the last "pure" refuge for "Germanness."[44] Taking a different perspective, environmentalists believe the return of the garden is symptomatic of the growing effects of globalization, climate change, and growing food insecurity.[45] The contemporary politics of capitalism, race, and food will only continue to frame debates over these marginal, beloved green spaces into the future.

In Germany, the garden pig may be out of fashion, but its relatives are making a comeback in other parts of Europe. In the United Kingdom, a new citizen-led project, known as the Pig Idea, has set the goal of feeding five thousand pigs annually within Greater London using the only food waste from restaurants and grocery stores.[46] The long-term goal is to reform the industrial food system by returning pigs to their traditional roles as recyclers. Germany and other EU countries may want to reconsider pigs as solutions to the problems of food waste. An opening for the garden pig will return, and when it does, Germans will find a working case study in histories of the GDR.

In late September 2018, Hurricane Florence drowned North Carolina in more than thirty inches of rain in four days. Almost immediately, pig manure was in the news and in North Carolina rivers. Since 1990, the state has become the second largest pork producer in the United States. Much of the pork industry is concentrated in the low-lying areas of the eastern half of the state.[47] The heavy rainfall overwhelmed manure lagoons that covered the region, drowning thousands of hogs and spilling millions of gallons of untreated waste water.[48] Many saw untreated animal waste and distended hog corpses as a warning from our future. Yet as *Communist Pigs* shows, this is not so new. In many ways, Hurricane Florence was a message from the past. As Walter Benjamin once wrote on the concept of history, "To articulate what is past does not mean to recognize 'how it really was.' It means to take control of a memory, as it flashes in a moment of danger."[49] North Carolina's flooded manure lagoons resonate with a memory of East German slurry ponds, a pulse of light, illuminating our warming world, then fading into shadow. In reminding us we've been here before, they demand we take control of that memory and act before it is too late. Though East Germany is long gone from the world, the world becomes more like East Germany every day.

NOTES

INTRODUCTION: ANIMAL FARMS

1 Orwell, *A Life in Letters*, 229.
2 Orwell and Davison, *George Orwell*, 229.
3 Arthur Schlesinger, "Mr. Orwell and the Communists," *New York Times*, August 25, 1946, section 7.
4 Rodden, *Understanding Animal Farm*, 146.
5 Saunders, *The Cultural Cold War*, 247–48; Adam Gopnik, "Honest, Decent, Wrong."

Animal Farm wasn't just reinterpreted by the US government; the CIA actually changed the ending. In 1951, future Watergate burglar and early Cold War CIA operative E. Howard Hunt purchased the film rights to the story from Orwell's widow, Sonia Orwell, in exchange for a promised introduction to Clark Gable. After CIA-hired psychologists scrutinized an early draft of the script, remarking that "there is no great clarity of message," the CIA changed the ending. The subsequent animated film, made in the United Kingdom in the 1950s, omitted Orwell's explicit characterization of despotic pigs as identical with their erstwhile capitalist enemies. Instead, the cartoon depicts the other livestock storming the farmhouse Bastille of the pigs, to begin the animal revolution all over again.

6 Rodden, *Understanding Animal Farm*, 142.
7 Orwell, *Animal Farm*, 155.
8 The question of inferiority, however, is a Western-centric one. See Siegelbaum, *The Socialist Car*; and Rubin, "Understanding a Car in the Context of a System German Socialism," 41.
9 After the first free elections of 1990 in the GDR, Socialist Democratic Party politician Otto Schily performed this perception on live television. When asked to explain the Central Democratic Union and Helmut Kohl's victory in the

East, Schily pulled a banana out of his pocket. Hellmuth Karasek, "Mit Kanonen auf Bananen?" *Der Spiegel* 13, March 26, 1990, 56–57; Timothy and Taylor, *Banana Wars*, 27.

10 Dale, *The East German Revolution of 1989*; Huff, *Natur und Industrie im Sozialismus*; Kallabis, *Ade, DDR!*; Maier, Dissolution; Neubert and Auerbach, *"Es kann anders werden"*; Ratzke, *Dagewesen und Aufgeschrieben*; Remy, *Opposition und Verweigerung in Nordthüringen*; Stoltzfus, "Public Space and the Dynamics of Environmental Action," 385–403.

11 Irene Jung, "Die Umwelt trägt die Hypotheken," *Hamburger Abendblatt*, August 2, 1990, www.abendblatt.de/archiv/1990/article202316553/Die-Umwelt-traegt-die-Hypotheken.html.

12 Dominick, "Capitalism, Communism, and Environmental Protection," 311.

13 Nelson, *Cold War Ecology*, xii.

14 Here I am borrowing from Weis's concept of "meatification," which I will explore in greater detail in chapters 2 and 3. Weis, *The Ecological Hoofprint*.

15 There are a number of great books on the cultural, economic, and biological history of pigs, including Essig, *Lesser Beasts*; Estabrook, *Pig Tales*; Gibson, *Feral Animals in the American South*; Malcolmson and Mastoris, *The English Pig;* Porter, *Pigs*; Rosenberg, "A Race Suicide among the Hogs"; and Watson, *The Whole Hog*; White, "From Globalized Pig Breeds to Capitalist Pigs," 94–120.

16 Larson, "Worldwide Phylogeography of Wild Boar," 1618–21. There is no single domestication event, but rather multiple sites of domestication.

17 Animal history/studies is a growing field largely dedicated to this fact. Here are just a few of the best examples: Greene, *Horses at Work*; Nance, *The Historical Animal*; Anderson, *Creatures of Empire*; Melville, *A Plague of Sheep*; Ritvo, *The Animal Estate*; Bulliet, *The Camel and the Wheel*; Taylor, *Making Salmon*; Gibson, *Feral Animals;* and Walker, *The Lost Wolves of Japan*.

18 Laue, *Das sozialistische Tier.*

19 Dale, *Between State Capitalism and Globalisation*, 195.

20 Kotkin, "The Kiss of Debt.".

21 Moore, *Capitalism in the Web of Life*; Patel, "The Long Green Revolution"; and Patel and Moore, *A History of the World in Seven Cheap Things*. Moore and Patel's theory provides the central analytical framework of this book, which I return to repeatedly. Their work argues that capitalism is a way of organizing nature into "cheap" things. Over the past five hundred years, it has constantly remade planetary nature to suit the expansion and accumulation of capital. Their "capitalism-as-world-ecology" theory argues that the separation of nature and culture was foundational to the emergence of capitalism as a historical force. This separation created the category of so-called "free gifts" from nature, which Moore and Patel identify as food, energy, work, care, money, raw materials, and lives, which were extracted from the "frontiers" of capitalist expansion. These frontiers, however, were not merely spatial. They were also biophysical, material, and social. Since the late fifteenth century,

capitalism has extracted cheap nature from a variety of such frontiers, including colonies, but also bodies (slaves' and women's reproduction), the subterranean world (coal and crude oil), and the atmosphere (synthetic nitrogen).

22 Poutrus, *Die Erfindung des Goldbroilers.* I am indebted to Poutrus's book for uncovering connections between East German agriculture, the chicken factory farm, and the GDR's international relationships.

23 Dominick, "Capitalism, Communism, and Environmental Protection," 315.

24 Laue, *Das Sozialistische Tier*; Poutrus, *Die Erfindung des Goldbroilers.* These works are built upon a generation of research into the nature of the East German dictatorship and society. Some of the most salient works include Fulbrook, *Anatomy of a Dictatorship*; Fulbrook, *The People's State*; Pence and Betts, *Socialist Modern*; and Ross, *Constructing Socialism at the Grass-Roots.* For an excellent, comprehensive overview of the East German historiography on socialist modernity, see the introductions to Moranda, *The People's Own Landscape*; and Rubin, *Synthetic Socialism.*

25 Orwell, *A Life in Letters*, 236.

26 Watson, *The Whole Hog.*

27 White, "Globalized Pig Breeds."

28 White, "Globalized Pig Breeds," 94–120; Porter, *Pigs*, 83.

29 Watson, *The Whole Hog*, 102.

30 See Haraway, *Simians, Cyborgs, and Women*; Latour, *The Pasteurization of France*; Latour, *Pandora's Hope*; and Whatmore, *Hybrid Geographies.*

31 Ingram, "Fermentation, Rot, and Other Human-Microbial Performances," 101–3.

32 Haraway, *When Species Meet.* As Haraway points out, the distinction between humans and nature is easily upended by the myriad microorganisms that colonize our guts.

33 Anderson, *Capitalist Pigs.*

34 Saraiva, *Fascist Pigs*, 131.

35 While not specifically focused on pigs, Rebecca Woods and Joshua Specht use animals to trace the expansion of British and American Empire, respectively. Woods, *The Herds Shot Round the World*; Specht, *Red Meat Republic.*

36 Moore, *Capitalism in the Web of Life*; Patel, "Long Green Revolution"; Patel and Moore, *Seven Cheap Things.*

37 Fitzgerald, *Every Farm a Factory*; Hurt, *Problems of Plenty*; Rosenberg, *The 4-H Harvest.*

38 Worster, *Dust Bowl.*

39 Here I am drawing from the insights of Gareth Dale, who builds on the work of Giovanni Arrighi and Tony Cliff to define the GDR as state capitalist. Dale, *Between State Capitalism and Globalisation.*

40 Haney, *When Modern Was Green*; Meyer-Renschhausen and Holl, *Die Wiederkehr der Gärten*; Willes, *The Gardens of the British Working Class.*

CHAPTER ONE: WHEN PIGS COULD FLY

1. "Schweine flogen durch die Luft," *agrar-PROFIL*: Organ der Parteileitung der BPO der SED im Komplex der Industriemäßigen Fleischproduktion Eberswalde, October 6, 1989, 8.
2. Porter, *Pigs*, 26.
3. "Schweine flogen durch die Luft," *agrar-PROFIL*, 8.
4. Holden and Ensminger, *Swine Science*, 22.
5. Blanchette, "Herding Species," 640–69.
6. Unger, *Der Berliner Bär*, 11.
7. "Agrarwirtschaft," Gesellschaft zur Erforschung und Förderungder Märkischen Eiszeitstraße e.V., http://wirtschaftsgeschichte-eberswalde.de/agrarwirtschaft/veb-schlacht-und-verarbeitungskombinat-eberswaldebritz-svke. The information here is partially provided through an online exhibit on the economic history of Eberswalde during the GDR.
8. BArch, DK 1/24582, p. 2. Schlacht- und Verarbeitungskomplex Eberswalde, Vertrag für ein "Schlüsselvertiges" Schlacht- und Verarbeitungskombinat, 1975.
9. "Bildung/Sozialpolitik," *agrar-PROFIL* 6 (October 1989): 7.
10. "Production, Supply and Distribution Online," Foreign Agricultural Service, United States Department of Agriculture, International Pork Production Statistics, www.fas.usda.gov/psdonline/psdQuery.aspx.
11. Nolan, *The Transatlantic Century*, 191.
12. Schneider, *The Wall Jumper*, 119.
13. Brown, "Gridded Lives," 17–48.
14. Fitzgerald, *Every Farm a Factory*.
15. Fitzgerald, *Every Farm a Factory*, 109.
16. This is a story recounted in many places, but perhaps done best by Fitzgerald, *Every Farm a Factory*; Gilbert, *Planning Democracy*; Hurt, *Problems of Plenty*; Rosenberg, *The 4-H Harvest*, 5, 58; and Stoll, *The Fruits of Natural Advantage*.
17. Nolan, *Visions of Modernity*.
18. Nolan, *Visions of Modernity*, 61–64.
19. Nolan, *Visions of Modernity*, 65.
20. Rosenberg, *The 4-H Harvest*, 60.
21. Blackbourn, *The Conquest of Nature*, 63; Finlay, "The German Agricultural Experiment Stations," 46–50; Zimmerman, *Alabama in Africa*, 70–73.
22. Uekötter, *Die Wahrheit ist auf dem Feld: Eine Wissensgeschichte der deutschen Landwirtschaft*, 277–78. Germany's most successful tractor of this period, the Lanz Bulldog, was the world's first crude-oil-burning tractor, and it shared a great deal with the tractors of its American competitor, John Deere. Not only had the company founder, Heinrich Lanz, met John Deere in 1902, but later models, including the 1931 Lanz Bulldog HN1, were based on a contemporary John Deere model. The American manufacturer loved Lanz tractors so much that it bought the firm in 1956 to serve as its European arm. Don Macmillan, *The Big Book of John Deere Tractors*, 26, 76.

23 Tooze, *The Wages of Destruction*, 9–10.
24 Tooze, *The Wages of Destruction*, 470.
25 Saraiva, *Fascist Pigs*, 162.
26 Saraiva, *Fascist Pigs*, 158; Evans, *The Third Reich at War*, 295–303.
27 Fitzgerald, *Every Farm a Factory*, 157–83.
28 Patel and Moore, *Seven Cheap Things*, 1–42.
29 Bibliothek Berlin, 96 C 450a. Ulli Wruck, Offener Brief an alle Genossenschaftsbauern in den LPG Typ I des Bezirkes Neubrandenburg/von Ulli Wruck, Vorsitzender, und Heinz Behm, Feldbaubrigadier der LPG Typ 1 in Passow, Kreis Demmin (Neubrandenburg: Ideologische Kommission der SED-Bezirksleitung, 1963).
30 Worster, *Dust Bowl*, 45.
31 Harrison, *Driving the Soviets up the Wall*, 18.
32 Beyme, "Reconstruction in the German Democratic Republic," 193.
33 Nolan, *Transatlantic Century*, 177.
34 Harrison, *Driving the Soviets up the Wall*, 18.
35 Judt, *Postwar*, 16–21.
36 Naimark, *The Russians in Germany*; Slaveski, *The Soviet Occupation of Germany*.
37 Buechler and Buechler, *Contesting Agriculture*, 45; Nelson, *Cold War Ecology*, 61–63.
38 Groschoff and Heinrich, *Die Landwirtschaft der DDR*, 36.
39 Slaveski, *The Soviet Occupation of Germany*, 92; Tooze, *Wages*, 188.
40 For the Cold War period, I translate the German word *Bauern* as farmer rather than its traditional meaning of peasant, as it communicates to an English-speaking audience a more accurate description of the nature of their work.
41 Buechler and Buechler, *Contesting Agriculture*, 45.
42 Buechler and Buechler, *Contesting Agriculture*, 46.
43 Cullather, *The Hungry World*, 97.
44 Judt, *Postwar*, 105; Taylor, *Exorcising Hitler*, 122–23.
45 Cullather, *Hungry World*, 101.
46 Cullather, *Hungry World*, 97.
47 White, *The Organic Machine*; Worster, *Rivers of Empire*.
48 Cullather, *Hungry World*, 108–10.
49 Hurt, *Problems*, 118.
50 Walt Rostow, "Marx Was a City Boy or, Why Communism May Fail," *Harper's Magazine*, February, 1955. https://harpers.org/archive/1955/02/marx-was-a-city-boy/6.
51 Corey, *Meat and Man*, 341.
52 Weis, *The Ecological Hoofprint*. For a more in-depth discussion of Weis's work, see chapter 2.
53 Of course, other historians have puzzled over the ways communist and capitalist state power have remade space in remarkably similar ways. See

Brown, "Gridded Lives"; Brown, *Plutopia*; Scott, *Seeing like a State*; and Zeisler-Vralsted, *Rivers, Memory, and Nation-Building.*

54 Steiner, *The Plans That Failed*, 76. Soviet administrators made some attempts to ameliorate the poor situation, giving the 210,000 new farmers another three hectares of land each in 1950.

55 Buechler and Buechler, *Contesting Agriculture*, 46. Depending on the type of LPG, farmers received a different percentage of the income from pooled land. In Type I, members were entitled to 40 percent of the yield relative to the amount property they donated, while Type II and Type III received 30 percent and 20 percent, respectively. The state employed both hard and soft power to get farmers to join cooperatives.

56 Buechler and Buechler, *Contesting Agriculture*, 46.

57 Buechler and Buechler, *Contesting Agriculture*, 47, 90–99.

58 Port, *Conflict and Stability in the German Democratic Republic*, 71.

59 Bauerkämper, "Abweichendes Verhalten in der Diktatur," 298.

60 Bauerkämper, "Abweichendes Verhalten in der Diktatur," 76.

61 Bauerkämper, "Abweichendes Verhalten in der Diktatur," 299–300.

62 Osmond, "From Junker Estate to Co-operative Farm," 144–45.

63 Buechler and Buechler, *Contesting Agriculture*, 47–48.

64 Laue, *Das Sozialistische Tier*, 161–65.

65 Bauerkämper, "Abweichendes Verhalten," 300.

66 Osmond, "Junker Estate," 144.

67 Port, *Conflict and Stability in the German Democratic Republic*, 88.

68 Port, *Conflict and Stability in the German Democratic Republic*, 90–93.

69 Hamilton, *Trucking Country*, 19; Gilbert, *Planning Democracy.*

70 Hamilton, *Trucking Country*, 114–17; Hurt, *Problems*, 116.

71 Hurt, *Problems*, 117.

72 Hamilton, *Trucking Country*, 101.

73 Hamilton, *Trucking Country*, 114–17.

74 Hamilton, "Agribusiness," 576. France's postwar modernization drive offers another intereresting parallel to this shared history of industrial agriculture. See Bivar, *Organic Resistance*, 13–47.

75 Hamilton, "Agribusiness," 577.

76 Hamilton, "Agribusiness," 578.

77 Hurt, *Problems*, 117.

78 Anderson, *Industrializing the Corn Belt*, 5. According to the *Des Moines Register*, the decline has continued unabated into the more recent past. Donnelle Eller and Christopher Doering, "Ag Census Finds Iowa Farms Are Bigger But Fewer in Number," *Des Moines Register*, February 20, 2014, www.desmoinesregister.com/story/money/agriculture/2014/02/21/ag-census-finds-iowa-farms-are-bigger-but-fewer-in-number/5669313; *1969 Census of Agriculture: Area Reports. Iowa* (US Government Printing Office, 1972), 3.

79 Hamilton, *Trucking Country*, 100.

80 Fitzgerald, "Eating and Remembering," 395–96, 404. Fitzgerald quotes here from a study produced by the Leopold Society for Sustainable Agriculture at Iowa State University.

81 Hurt, *Problems*, 117.

82 Gary Krob, "Iowa's Changing Demographics," accessed May 1, 2018, www.legis.iowa.gov/docs/publications/SI/794317.pdf.

83 Trager, *Amber Waves of Grain*, 93.

84 Anderson, *Corn Belt*, 120.

85 Trager, *Amber Waves of Grain*, 93.

86 Nelson, *Farm and Factory*, 179.

87 In the 1980s, farmer suicide became national news as rates in Iowa climbed well above the national average. Keith Schneider, "Rash of Suicides in Oklahoma Shows that the Crisis on the Farm Goes On," *New York Times*, August 17, 1987; Berry, *The Unsettling of America*, 3; Osha Gray Davidson, *Broken Heartland*, 94–97.

88 Berry, *The Unsettling of America*, 41.

89 Berry, *The Unsettling of* America, 61.

90 Josephson, *Would Trotsky Wear a Bluetooth?*, 3–5.

91 Steiner, *Plans*, 90–91.

92 Kopstein, *The Politics of Economic Decline in East Germany*, 64, 90–91.

93 Harrison, *Driving the Soviets up the Wall.*

94 Kopstein, *Economic Decline*, 64; Poutrus, *Die Erfindung des Goldbroilers*, 29–35; Spufford, *Red Plenty*; Trager, *Amber Waves.*

95 Poutrus, *Goldbroilers*, 85.

96 Selverstone, *Constructing the Monolith*, 116–44.

97 Lampe, Prickett, and Adamović, *Yugoslav-American Economic Relations*, 81.

98 Lampe, Prickett, and Adamović, *Yugoslav-American Economic Relations*, 117.

99 Hamilton, "Supermarkets USA Confronts State Socialism."

100 Poutrus, *Goldbroilers*, 86.

101 Lampe, Prickett, and Adamović, *Yugoslav-American Economic Relations*, 118.

102 Lampe, Prickett, and Adamović, *Yugoslav-American Economic Relations*, 84.

103 Poutrus, *Goldbroilers*, 86–88; 97–98.

104 Liselo Thoms and Peter Lorf, "Die Sozialistische Kooperation Stärkt die Wirtschaft Unserer Beider Länder, *Neues Deutschland* (October 1, 1966): 3.

105 "Ein ereignisreicher Tag in Slowenien," *Neues Deutschland*, September 30, 1966, 3.

106 Liselo Thoms and Peter Lorf, "Die Sozialistische Kooperation Stärkt die Wirtschaft Unserer Beider Länder, *Neues Deutschland*, October 1, 1966, 3.

107 Liselo Thoms and Peter Lorf, "Die Sozialistische Kooperation Stärkt die Wirtschaft Unserer Beider Länder, *Neues Deutschland*, October 1, 1966, 3.

108 Poutrus, *Goldbroilers*, 86–88.

109 Poutrus, *Goldbroilers*, 87.

110 Rudi Peschel, "Eier und Schinken wie vom Fließband," *Neues Deutschland*, November 29, 1967, 7.

111 Peschel, "Eier und Schinken wie vom Fließband," *Neues Deutschland*, November 29, 1967, 7.
112 Ritze, *Schweine*, 332; 375–76.
113 BArch, DC 14/162. Aufbau von Beispielanlagen für Industriemäßige Produktion in der Landwirtschaft, Band 1 (1968–69). Report, "Zur Überprüfung der Beispielanlage der Schweinezucht und Mast KIM Eberswalde," 1–18.
114 "So Fing Alles An," *agrar-PROFIL* 6 (October 1989): 4.
115 BArch, DC 14/162. Aufbau von Beispielanlagen für Industriemäßige Produktion in der Landwirtschaft, Band 1 (1968–69). Report, "Zur Überprüfung der Beispielanlage der Schweinezucht und Mast KIM Eberswalde," 10.
116 "Schweine fliegen durch den Luft," *agrar-PROFIL* 6 (October, 1989): 8.
117 Ibid.; "So Fing Alles An," 4.
118 Finlay, "Hogs, Antibiotics, and the Industrial Environments of Postwar," ix, 275.
119 Taiganides and International Development Research Centre, *Pig Waste Management*, 41–44.
120 Finlay, "Hogs, Antibiotics," 247.
121 Terent'eva, *Zu Einigen Fragen der Intensivierung der Industriemässigen Schweineproduktion*, 7, 20.
122 Porter, *Pigs*, 23–25.
123 Nitzsche and Paulke, *100 Jahre Schweinezuchtverband Brandenburg*, 25; Porter, *Pigs*, 23–25, 79; Terent'eva, *Intensivierung der Schweineproduktion*, 24. In that same year, the East Germans founded the VVB Schweinezucht, a cooperative operation for pig breeding.
124 Terent'eva, *Intensivierung der Schweineproduktion*, 79.
125 Smith, *Works in Progress*; Terent'eva, 21. Smith explains the history of herd book keeping in the Soviet Union.
126 Ensminger, *Swine Science*, 24; Terent'eva, *Intensivierung der Schweineproduktion*, 17.
127 Anderson, "Lard to Lean," 32–33.
128 Ritze et al., *Schweine: Zucht, Haltung, Fütterung*, 76.
129 Ritze et al., *Schweine: Zucht, Haltung, Fütterung*, 76.
130 In addition to the Leicoma, East German breeders established another hybrid pig, the Schwerfurter, the result of collaboration between facilities in the regions of Schwerin and Erfurt.
131 Gunther Nitzsche, "Zur Geschichte der Schweinerasse Leicoma," (paper presented at the conference of Leicoma Breeding Commission, Bad Sulza, 2006); Nitzsche and Paulke, *100 Jahre*, 40.
132 Ahrens, "East German Foreign Trade in the Honecker Years," x, 249; Smith, *Works in Progress*, 113; Terent'eva, *Intensivierung Der Schweineproduktion*, 23–25. As Smith shows, the Soviet Union did not keep complete breeding records until well after the Second World War.
133 Nitzsche, "Zur Geschichte der Schweinerasse Leicoma."
134 SAPMO BArch, DY/30/IV B 2/2.023/7, pp. 5–6. "Report on Problems of Pig Breeding," 27.10.76.

135 Peschel, "Eier und Schinken wie vom Fließband," *Neues Deutschland*, November 29, 1967, 7.
136 Kopstein, *Economic Decline*, 68.
137 Kopstein, *Economic Decline*, 86.
138 Peschel, "Eier und Schinken wie vom Fließband," *Neues Deutschland*, November 29, 1967, 7.
139 Peschel, "Eier und Schinken wie vom Fließband," *Neues Deutschland*, November 29, 1967, 7.

CHAPTER TWO: THE GREAT GRAIN ROBBERY AND THE RISE OF A GLOBAL ANIMAL FARM

1 Trager, *Amber Waves of Grain*, 83. My main source for the events surrounding the Great Grain Robbery is Trager's book.
2 Trager, *Amber Waves of Grain*, 23.
3 Luttrell, "The Russian Wheat Deal," 1.
4 Lutrell, "The Russian Wheat Deal," 2; Morgan, *Merchants of Grain*, 110–30.
5 Trager, *Amber Waves of Grain*, 1–2.
6 Trager, *Amber Waves of Grain*, 1–2.
7 Trager, *Amber Waves of Grain*, 186.
8 Trager, *Amber Waves of Grain*, 65.
9 Michael C. Jensen, "Soviet Grain Deal Is Called a Coup," *New York Times*, September 29, 1972, 66.
10 Weis, *The Ecological Hoofprint*, 8.
11 Weis, *The Ecological Hoofprint*, 8.
12 Nolan, *The Transatlantic Century*, 173–74.
13 Like in the introduction, I am referencing again Patel and Moore, *A History of the World in Seven Cheap Things*.
14 Trager, *Amber Waves of Grain*, 186.
15 While pivotal, 1972 marked only one year across nearly a decade of negotiations in the history of détente. Most historians mark the period as running from 1966 to the end of the 1970s.
16 Suri, *Power and Protest*. In this case, Suri largely attributes the impetus of the Moscow Summit to Nixon's visit to China, without mentioning the USDA delegation's visit in April.
17 Multiple independent targeted reentry vehicles, a type of nuclear missile that could release from a single rocket several separate nuclear warheads, like the Minutemen III.
18 Sachse, "Bullen, Hengste, Wissenschaftler"; Carlson, *K Blows Top*.
19 Trager, *Amber Waves of Grain*, 122.
20 Trager, *Amber Waves of Grain*, 11.
21 Schwartz, "Legacies of Detente," 517.
22 Trager, *Amber Waves of Grain*, 12–13.
23 Cullather, *The Hungry World*, 232.

24 Nove, "Will Russia Ever Feed Itself," 9.
25 Cullather, *Hungry World*, 7–15; Fitzgerald, *Every Farm a Factory*.
26 Carlson, *K Blows Top*.
27 Nove, "Will Russia Ever Feed Itself," 9; Trager, *Amber Waves of Grain*, 7.
28 While Borlaug's "miracle seeds," like new hybrids varieties of rice and wheat, garnered international attention, chemical fertilizers, expensive machinery, and disciplined labor made the record-breaking yields possible.
29 Nove, "Will Russia Ever Feed Itself," 9.
30 Agra Europe Special Report No. 52, etc. Agra Europe (London) ltd., 1990, 4.
31 BArch, DL 102/1372, 24.
32 Trager, *Amber Waves of Grain*, 9.
33 Essig, *Lesser Beasts*, 80–88; White, "From Globalized Pig Breeds to Capitalist Pigs," 94–120.
34 Porter, *Pigs*, 86–87; White, "From Globalized Pig Breeds to Capitalist Pigs," 97.
35 A Mrs. Houston, British visitor to Cincinnati in the 1830s, described the Queen City as "literally speaking, a *city of hogs* . . . a monster piggery. . . . Alive and dead, whole and divided into portions, their outsides and their insides, their grunts and their squeals, meet you at every moment." Richard Arms, "From Disassembly to Assembly: Cincinnati, the Birthplace of Mass-Production," *Bulletin of the Historical and Philosophical Society of Ohio* 92, no. 3, July, 1959, 199.
36 Cronon, *Nature's Metropolis*, 228–29; White, "Globalized Pig Breeds," 110.
37 Ensminger, *Swine Science*, 44–54.
38 Saraiva, *Fascist Pigs*, 110.
39 Saraiva, *Fascist Pigs*, 111.
40 Holden and Ensminger, *Swine Science*, 37–38.
41 Saraiva, *Fascist Pigs*, 111, 121–31.
42 EC regulations, import levies, and preferential access agreements, however, shut the Americans out.
43 Zimmerman, "Unraveling the Ties That Really Bind," 136.
44 Luttrell, "The Russian Wheat Deal," 2; Morgan, *Merchants*.
45 Special to the New York Times, "Cold Snap Causes Rail and Heating Problems," *New York Times*, January 9, 1973, A1.
46 Trager, *Amber Waves of Grain*, 83.
47 Helleiner, *Forgotten Foundations of Bretton Woods*, 40–45; Nolan, *Transatlantic Century*, 178.
48 These works offer excellent, succinct summaries of the Bretton Woods system: Frieden, *Global Capitalism*; Judt, *Postwar*; Nolan, *Transatlantic Century*; and Steil, *The Battle of Bretton Woods*.
49 Frieden, *Global Capitalism*, 343–45.
50 Frieden, *Global Capitalism*, 343–45. The sheer amount of cash involved put the stability of national currencies in play. American officials worried that a run by foreign banks and the financial industry on a clearly overvalued dollar would throw the American economy into chaos.

51 Dale, *Between State Capitalism and Globalisation*, 192–93. Deregulation, however, weakened the ability of the federal government to control or protect US financial markets over the long run.

52 Stein, *Pivotal Decade*.

53 Trager, *Amber Waves of Grain*, 211.

54 Nolan, *Transatlantic Century*, 285.

55 Judt, *Postwar*, 455.

56 Nolan, *Transatlantic Century*, 209. "In 1961, the Soviet Union provided 80 percent of Iceland's oil, 35 percent of Finland's, and between 19–22 percent of oil for Greece, Italy, Austria, and Sweden."

57 Stokes, "East Germany and the Oil Crises of the 1970s," 134–36.

58 Dale, *State Capitalism and Globalisation*, 197; Steiner, *The Plans That Failed*, 73.

59 Stokes, "East Germany and the Oil Crises of the 1970s," 134–37.

60 Ahrens, "East German Foreign Trade in the Honecker Years," 172.

61 Dale, *State Capitalism and Globalisation*, 197; Steiner, *The Plans That Failed*, 73.

62 Nolan, *Transatlantic Century*, 286.

63 Adelman, "International Finance and Political Legitimacy," 117.

64 Frieden, *Global Capitalism*, 370.

65 Weis, *Hoofprint*, 74.

66 Dale, *State Capitalism and Globalisation*, 196.

67 Zatlin, *The Currency of Socialism*, 70.

68 Frieden, *Global Capitalism*, 368.

69 Dale, *State Capitalism and Globalisation*, 195.

70 Dale, *State Capitalism and Globalisation*, 194.

71 Last, *After the "Socialist Spring,"* 81; Zatlin, *The Currency of Socialism*.

72 Burawoy and Lukács, *The Radiant Past*.

73 Kopstein, *Economic Decline*, 46–60; Zatlin, *The Currency of Socialism*, 48–50; Stokes, *Constructing Socialism*, 46–60.

74 Steiner, *The Plans that Failed*, 143.

75 Judt, *Postwar*, 498.

76 Dale, *State Capitalism and Globalisation*, 179.

77 Dale, *State Capitalism and Globalisation*, 187–89.

78 Karlsch, "National Socialist Autarky Projects," 92.

79 Jensen, "Soviet Grain Deal Is Called a Coup," 66.

80 Weis, *Hoofprint*, 8, 58–70.

CHAPTER THREE: THE SHRINKING INDUSTRIAL PIG

1 SAPMO BArch, DY 30/1741. Abteilung Landwirtschaft im ZK der SED—Natur- und Umweltschutz. Report, August 18, 1980, 13.

2 Manfred Grund, "Aus Jeder Besamung Drei Ferkel Mehr," *Neue Deutsche Bauernzeitung*, August 6, 1982, 8–9.

3 Manfred Grund, "Aus Jeder Besamung Drei Ferkel Mehr," *Neue Deutsche Bauernzeitung*, August 6, 1982, 8–9.

4 Manfred Grund, "Aus Jeder Besamung Drei Ferkel Mehr," *Neue Deutsche Bauernzeitung*, August 6, 1982, 8–9.

5 SAPMO BArch, DY30/IV B 2/2.023/34. Development of Farms along Industrial Principles (1975). Memo from Grüneberg titled "Problems for Discussion with Comrade Schürer on Monday, April 21, 1975," 8–9.

6 Schier, *Alltagsleben im "Sozialistischen Dorf,"* 145.

7 Last, *After the "Socialist Spring,"* 145.

8 Jacobsen, "The Foreign Trade and Payments of the GDR in a Changing World Economy," 240–41.

9 SAPMO BArch, DY30/IV B 2/2.023/34. Development of Farms along Industrial Principles (1975). Heinz Kuhrig letter draft for Grüneberg to Erich Honecker on the question of "welchen zusätzlichen gesamtvolkswirtschaftlichen Beitrag die Land und Nahrungsgüterwirtschaft mit der Plandurchführung leisten könnte," January 20, 1975.

10 Agra Europe Special Report No. 52, The Agricultural Industry of the German Democratic Republic (London: AGRA EUROPE LTD, 1990), 18.

11 SAPMO BArch, DY30/IV B 2/2.023/34. Development of Farms along Industrial Principles (1975). Correspondence between Grüneberg and Honecker, March 13, 1975

12 BArch, DK 108–9, Band 1. Newspaper clipping, Egbert Steinke, "Der Erfolg kam aus dem Ostsen: Berlin-Consult erhielt Großaufträge aus Polen und der DDR," *Handelsblatt Deutsche Wirtschaftszeitung*, January 8, 1972. The West German agricultural investment firm was Institut Strukturforschung und Planung.

13 BArch, DK 1/24582. Contract between Berlin Consult and VE Außenhandelsb-etrieb Industrieanlage-Import, February 20, 1975.

14 BArch, DK 108–9, Band 1. Newspaper clipping, "Die Fleischerei: Größte Fleischfabrik Europas Wird an der DDR Geliefert," no. 5, May 1975, BRD, 54.

15 Ahrens, "East German Foreign Trade in the Honecker Years," 168.

16 Kopstein, *The Politics of Economic Decline in East Germany*, 82–86.

17 Steiner, *The Plans That Failed*, 161.

18 SAPMO BArch, DY30/IV B 2/2.023/34. Development of Farms along Industrial Principles (1975). Letter from Schürer and Mittag to Honecker, February 26, 1975; Kopstein, *Economic Decline*, 82–86.

19 Last, *After the Socialist Spring*, 158.

20 SAPMO BArch, DY30/IV B 2/2.023/34. Development of Farms along Industrial Principles (1975). Grüneberg Writing to Honecker on March 13, 1975, pp. 1–2.

21 SAPMO BArch, DY30/IV B 2/2.023/34. Development of Farms along Industrial Principles (1975). Memo from Grüneberg to Honecker on March 13, 1975, p. 2.

22 SAPMO BArch, DY30 IV B 2/2.023/34. Development of Farms along Industrial Principles (1975). Report titled "Notes from the Deliberation of Comrade Schürer with the MLFN about the Plan of 1976 and Basic Problems of 1976–80 on April 1, 1975," April 4, 1975, p. 4.

23 SAPMO BArch, DY30 IV B 2/2.023/34. Development of Farms along Industrial Principles (1975). Report titled "Notes from the Deliberation of Comrade Schürer with the MLFN about the Plan of 1976 and Basic Problems of 1976–80 on April 1, 1975," April 4, 1975, p. 4.

24 SAPMO BArch, DY30/IV B 2/2.023/34. Development of Farms along Industrial Principles (1975). Memo from Grüneberg titled "Problems for the Debate with Comrade Schürer on Monday, April 21, 1975," p. 1.

25 SAPMO BArch, DY30/IV B 2/2.023/34. Development of Farms along Industrial Principles (1975). Memo from Grüneberg titled "Problems for the Debate with Comrade Schürer on Monday, April 21, 1975."

26 SAPMO BArch DY30/IV B 2/2.023/34. Development of Farms along Industrial Principles (1975). Heinz Kuhrig to Gerhard Grüneberg, April 16, 1975, Cover Letter to study "Socialist Intensification and the Transition to Industrial Production," pp. 1–2; Buechler and Buechler, *Contesting Agriculture*, 61–62; Last, *After the Socialist Spring*, 181–83. As Last writes, the regime had begun experimenting with hyper specialization in 1971 with the creation of Cooperative Units for Grain Production (or KAPs)—a transitory property arrangement between collective and state property, which anticipated the LPG Ps. The KAPs were made up of land contributed by several neighboring LPGs for the production of grain and often fodder. This produce was then sent on to state-owned livestock facilities and food-processing factories or exported directly for international trade.

27 SAPMO BArch, DY30/IV B 2/2.023/34. Development of Farms along Industrial Principles (1975). Heinz Kuhrig to Gerhard Grüneberg, April 16, 1975, cover letter to study of "Socialist Intensification and the Transition to Industrial Production," p. 3.

28 SAPMO BArch, DY30/IV B 2/2.023/34. Development of Farms along Industrial Principles (1975). Heinz Kuhrig to Gerhard Grüneberg, April 16, 1975. Study, "Socialist Intensification and the Transition to Industrial Production," pp. 12–13; Editor, "Was bietet die agra 79?," *Neue Deutsche Bauernzeitung* 23 (June 8, 1979):16–17.

29 SAPMO BArch, DY30/IV B 2/2.023/34. Development of Farms along Industrial Principles (1975). Memo from Grüneberg titled "Problems for Discussion with Comrade Schürer on Monday, April 21, 1975," 5.

30 BArch DC, 20/20371, p. 1. "Increase in Fruit, Vegetable, and Fodder Productions" from the Working Group of the Ministers' Council for Organization and Inspection, 1977.

31 Groschoff and Heinrich, *Die Landwirtschaft der DDR*, 38.

32 Eli Rubin, *Synthetic Socialism;* Ulrich Petschow et al., *Umweltreport DDR*, 26–27. Brown coal was an especially toxic feed stock for the plastic industry. Its energy content was low and therefore burned inefficiently, producing toxic dust particles, which blackened concrete, killed trees, and clung to lungs and skin, while also leaving a high percentage of non-combustible material behind.

33 Erich Honecker, "Grußadresse des Zentralkomittees der SED zum Tag der Genossenschaftsbauern und Arbeit der Sozialistischen Land- und Forstwirtschaft der DDR," republished in *Neue Deutsche Baurenrzeitung* 26 (June 25, 1982): 2.
34 SAPMO BArchB, DY30 IV B 2/2.023/34. Development of Farms along Industrial Principles (1975). Memo from Grüneberg titled "Problems for Discussion with Comrade Schürer on Monday, April 21, 1975," p. 4.
35 Last, *After the Socialist Spring*, 187.
36 Holden and Ensminger, *Swine Science*, 130–31.
37 Holden and Ensminger, *Swine Science*, 23.
38 Watson, *The Whole Hog*, 99–100.
39 Watson, *The Whole* Hog, 99–101.
40 Groschoff and Heinrich, *Die Landwirtschaft der DDR*, 368.
41 Groschoff and Heinrich, *Die Landwirtschaft der DDR*, 368.
42 Holden and Ensminger, *Swine Science*, 163.
43 Merkel, "Agriculture," 233.
44 Jacobsen, "The Foreign Trade and Payments of the GDR in a Changing World Economy," 244.
45 BArch DC, 20/20371. "Measures for the Strengthened Use of Fruit, Vegetable, and Fodder Reserves in Villages and Cities," 7:25 (1977), 1.
46 SAPMO BArch, DY 30/IV B 2/2.023/35. Meat Supply and Foot and Mouth Disease Reports, 1975. Memo from Kuhrig to Grüneberg, "Development of Agriculture between 1976 and 1980," July 8, 1975, 1.
47 SAPMO BArch, DY 30/J IV 2/2/1764. Meeting of February 6, 1979. "Vorschläge zur Lösung des Getreide- und Futtermittelproblems."
48 Curry, "Atoms in Agriculture," 129–30; Hamblin, "Quickening Nature's Pulse," 389–408.
49 BArch DC, 20/20371, "Increase in Fruit, Vegetable, and Fodder Productions," from the Working Group of the Ministers' Council for Organization and Inspection, 1977. Attachment 4, "Long-term Measures for the Effective Use of Land and First Results in the Community Association of Liberose, Kreis Beeskow, Bezikrk Frankfurt/Oder, 1."
50 BArch DC, 20/20371. Increase in Fruit, Vegetable, and Fodder Productions (1977). Report, "Measures to Strengthen the Use of Reserves for the Production of Fruit, Vegetables, and Fodder in Villages and Cities," July 25, 1977, Attachment 4, "Long-term Measures for the Effective Use of Land and First Results in the Community Association of Liberose, Kreis Beeskow, Bezirk Frankfurt/Oder," 1. In an "exchange," individuals had their property rights preserved, but their land was swapped for a less "economic" plot on the farm. See Katherine Verdery, *The Vanishing Hectare*.
51 BArch DC, 20/20371. Increase in Fruit, Vegetable, and Fodder Productions, (1977). Report, "Measures to Strengthen the Use of Reserves for the Production of Fruit, Vegetables, and Fodder in Villages and Cities," July 25, 1977,

Attachment 4, "Long-term Measures for the Effective Use of Land and First Results in the Community Association of Liberose, Kreis Beeskow, Bezirk Frankfurt/Oder," 1.

52 Groschoff and Heinrich, *Die Landwirtschaft der DDR*, 327.

53 Schier, *Alltagsleben im "Sozialistischen Dorf."*

54 In an attempt to streamline decision making, the regime empowered district councils or *Kreisräte* to take a larger role in coordinating district wide production levels during the mid-1970s. Their power was not absolute as the LPGs maintained preeminent status, but they were made up of chairmen from other LPGs as well as representatives from agricultural academic institutions.

55 Schier, *Alltagsleben im "Sozialistischen Dorf,"* 164–65. The director's name was Rudi Bogner, and Schier gives him a prominent role in her narrative.

56 Schier, *Alltagsleben im "Sozialistischen Dorf,"* 167–68.

57 BArch DC, 20/20371. Increase in Fruit, Vegetable, and Fodder Productions, (1977). Report, "Measures to Strengthen the Use of Reserves for the Production of Fruit, Vegetables, and Fodder in Villages and Cities," July 25, 1977, pp. 2–3.

58 Herbert Weimar, "Ertrag von Jedem Quadratmeter Land," *Neue Deutsche Bauernzeitung*, 26 (June, 25, 1982): 5–6.

59 Weimar, "Ertrag von Jedem Quadratmeter Land," 5.

60 LAB C. Rep. 112 No. 276. Development of Livestock Production in Berlin (1975). "Toward the Development of Livestock in the Agriculture of Berlin," November 11, 1975, 3.

61 LAB C. Rep. 112 No. 276. Development of Livestock Production in Berlin (1975). "Toward the Development of Livestock in the Agriculture of Berlin," November 11, 1975, 3–4.

62 McNeur, *Taming Manhattan*.

63 Essig, *Lesser Beasts*, 203.

64 Fairlie, *Meat*, 48.

65 Anderson, *Capitalist Pigs*, 98–100. Uncooked organic waste is the most prevalent source of parasitic worms in pigs, which then pass them on to people in their meat.

66 Autorkollectiv, "Mit Resten Mästen," *Neue Deutsche Bauernzeitung* 34 (August 20, 1982): 8–9.

67 Autorkollectiv, "Mit Resten Mästen," *Neue Deutsche Bauernzeitung*, 34 (August 20, 1982): 8–9.

68 SAPMO BArch, DY 30/IV B 2/2.023/35. Meat Supply and Foot and Mouth Disease Reports (1975). "Report on Fertilizers," July 4, 1975, 2.

69 SAPMO BArch, DY 30/IV B 2/2.023/35. Meat Supply and Foot and Mouth Disease Reports (1975). "Report on Fertilizers," July 4, 1975, 3.

70 Soule and Piper, *Farming in Nature's Image*, 14.

71 BArch, DK 5/1730. Nitrates and Water Contamination (1979). Report, "Study of Nitrate Pollution in the Bodies of Water of GDR, Its Development over the Last

Few Years, and Main Measures for Change," Institute for Water, Berlin, February 1979. Summary, 4.

72 Günter Glowka, "Verluste durch Unterlassene Nitrattests," *Neue Deutsche Bauernzeitung* 36/1982 (September 3, 1982): 13.

73 Glowka, "Verluste durch Unterlassene Nitrattests," 13.

74 BArch, DK 5/1730, p. 86. Nitrates and Water Contamination (1979). Report, "Study of Nitrate Pollution in the Bodies of Water of GDR, Its Development over the Last Few Years, and Main Measures for Change," Institute for Water, Berlin, February 1979.

75 BArch, DK 5/1730, pp. 74–75. Nitrates and Water Contamination (1979). Report, "Study of Nitrate Pollution in the Bodies of Water of GDR, Its Development over the Last Few Years, and Main Measures for Change," Institute for Water, Berlin, February 1979. Nitrate concentration becomes incredibly dangerous when it exceeds 500mg/kg of freshly harvested fodder (like grasses) and 5000mg/kg of dry fodder (like concentrates).

76 BArch, DK 5/1730, p. 86. Nitrates and Water Contamination (1979). Report, "Study of Nitrate Pollution in the Bodies of Water of GDR, Its Development over the Last Few Years, And Main Measures for Change," Institute for Water, Berlin, February 1979.

77 BArch, DK 5/1730, 86. Nitrates and Water Contamination (1979). Report, "Study of Nitrate Pollution in the Bodies of Water of GDR, Its Development over the Last Few Years, and Main Measures for Change," Institute for Water, Berlin, February 1979.

78 Laue, *Das Sozialistische Tier*, 208.

79 Laue, *Das Sozialistische Tier*, 208.

80 "Plants Poisonous to Livestock—Saponins," Department of Animal Science, College of Agriculture and Life Sciences, Cornell University, last modified September 10, 2015, http://poisonousplants.ansci.cornell.edu/toxicagents/saponin.html.

81 Holden and Ensminger, *Swine Science*, 240–42.

82 LAB C Rep. 112/252. Petition Analysis for Food Production and Agriculture (1975–80), Area of Veterinary Medicine, Berlin, January 7, 1976. Petition Analysis for the Forth Quarter 1975, 1. In the same file, but from later in the year, July 5, 1976, petitions sent to the district veterinarian discussed, by and large, the poor quality of groceries, especially meat and its "overly fatty" content.

83 Autorkollektiv, "DDR-Landwirtschaft im Spiegel von Drei Jahrzehnten," *Neue Deutsche Bauernzeitung* 24 (June 15, 1979): 3.

84 Autorkollektiv, "Im Spiegel von Drei Jahrzehnten," 3.

85 Autorkollektiv, "Im Spiegel von Drei Jahrzehnten," 4.

86 SAPMO BArch, DY 30/J IV 2/2/1764, p. 162. Vorschläge zur Lösung der Getreide- und Futtermittelproblems bei Maximaler Senkung von Futtermittel-importen aus dem NSW im Zeitraum 1985; Jacobsen, "The Foreign Trade and Payments of the GDR," 244.

87 SAPMO BArch, DY 30/J IV 2/2 1764, p. 160. Vorschläge zur Lösung der Getreide- und Futtermittelproblems bei Maximaler Senkung von Futtermittelimporten aus dem NSW im Zeitraum 1985.

CHAPTER FOUR: THE MANURE CRISIS

1 *Heimat* translates roughly as homeland and refers to a nineteenth-century concept that rooted German national identity in a romantic vision of small towns and pastoral landscapes. Confino, *The Nation as a Local Metaphor.*
2 Fridger Pelta, "Vogtländische Dungverstecke," *Neue Deutsche Bauernzeitung*, June 4, 1982, 5.
3 Fridger Pelta, "Vogtländische Dungverstecke," *Neue Deutsche Bauernzeitung*, June 4, 1982, 5.
4 Fridger Pelta, "Vogtländische Dungverstecke," *Neue Deutsche Bauernzeitung*, June 4, 1982, 5.
5 BArch, DK 5/1730, p. 6. Ministry of Environmental Protection and Water Management. Nitrate and Water Contamination (1979). Institute of Water Management, "Studie zur Stickstoffbelastung der Gewässer der DDR," Berlin, February 1979. The internationally recognized limit has since been lowered to 10 mg/l today.
6 Augustine, *Taking on Technocracy*, 201.
7 Schwarzer, *Sozialistische Zentralplanwirtschaft in der SBZ/DDR*, 154; Laue, *Das Sozialistische Tier*, 203n55.
8 Schönfelder, *Mit Gott Gegen Gülle*, 204.
9 SAPMO BArch, DY 30/1689, p. 151. Monthly, Quarterly, and Annual Results for Livestock Production. "Planerfüllung Ergebnisse für die Abteilung Landwirtschaft des ZK, 3." March, 1982.
10 Edwards and Driscoll, "Environmental Consequences of Swine," 155–70.
11 Stoll, *Larding the Lean Earth*, 50–51.
12 Charles Fulhage and Joe Harner, "Liquid Manure Handling Systems," Cornell University Cooperative Extension, at eXtension.org, February 16, 2018, www.extension.org/pages/8905/liquid-manure-collection-and-handling-systems.
13 West and Kennedy, *Swine Manure Handling Systems.*
14 West and Kennedy, *Swine Manure Handling Systems*, 3–5; Edwards and Driscoll, "Environmental Consequences of Swine," 160.
15 Edwards and Driscoll, "Environmental Consequences of Swine," 162.
16 Holden and Ensminger, *Swine Science*, 23.
17 Edwards and Driscoll, "Environmental Consequences of Swine," 155–70.
18 SAPMO BArch, DY 30/1687, p. 233. Gülle und Tierhygiene (1975–79). "Information about the Progress of 'Bio-slurry Preparation' in SZMK Eberswalde," July 14, 1978. A twelve-kiloton facility refers to the total live weight of all the animals in the farm. Designating farms by the weight of their animals helped planners standardize comparisons between facilities. Animal units

(Großvieheinheiten, or GE) helped planners compare apples to oranges, so to speak. In this case, five pigs were equal to about one cow.

19 Griffith, *Husbandry*, 236–37.

20 SAPMO BArch, DY 30/1687, p. 233. Gülle und Tierhygiene (1975–79). Report, "Bio-slurry" at SZMK Eberswalde, July 14, 1978; West and Kennedy, *Swine Manure Handling Systems*, 36.

21 SAPMO BArch, DY 30/1687, p. 233. Gülle und Tierhygiene (1975–79). Report, "Progress of Work, Problems, and the New Developments in the Area of Preparation and Evaluation of Bio-slurry in SZMK Eberswalde," July 14, 1978.

22 SAPMO BArch, DY 30/1687, pp. 234–35. Gülle und Tierhygiene (1975–79). Report, "Progress of Work, Problems, and the New Developments in the Area of Preparation and Evaluation of Bio-slurry in SZMK Eberswalde," July 14, 1978.

23 SAPMO BArch, DY 30/1687, p. 236. Gülle und Tierhygiene (1975–79). Report, "Progress of Work, Problems, and the New Developments in the Area of Preparation and Evaluation of Bio-slurry in SZMK Eberswalde," July 14, 1978. From a section in the report entitled, "Futter-Eiweißgewinnung aus der Mikrobiologischen Behandlung von Schweinegülle"; Stoll, *Larding the Lean Earth*, 52.

24 SAPMO BArch, DY 30/1687, p. 236. Gülle und Tierhygiene (1975–79). Report, "Progress of Work, Problems, and the New Developments in the Area of Preparation and Evaluation of Bio-slurry in SZMK Eberswalde," July 14, 1978.

25 SAPMO BArch, DY 30/1577, pp. 351–52. Kombinat Industrielle Tierproduktion, Band 3 (1982–83). "Information zur Produktion und zum Export in das NSW von Trockenbiomasse für Düngungszwecke aus Überschußschlamm der Gülleaufarbeitung mit der Bezeichnung als Biodünggestoff," May, 1983.

26 BArch, DK, 5/3264. Water Management and Manure (1973). Letter from Dr. Reichelt, minister of environmental protection and water management, to one Genosse Ewald, May 7, 1973. Attached Resolution, under section titled "Guidelines for Water Usage in the Application of Manure to Grain Production."

27 SAPMO BArch DY, 30/1687, p. 27. Gülle und Tierhygiene (1975–79). Report on manure processing in the LPG Trinwillershagen, January 8, 1976.

28 SAPMO BArch DY 30/1687. p. 49. Gülle und Tierhygiene (1975–79). Report, "Update on Manure Handling Experimentation at SZMK Eberswalde, September 17, 1976."

29 SAPMO BArch DY 30/1687, p. 238. Gülle und Tierhygiene (1975–79). Report, "Progress of Work, Problems, and the New Developments in the Area of Preparation and Evaluation of Bio-slurry in SZMK Eberswalde," July 14, 1978.

30 SAPMO BArch, DY 30/1687, pp. 187–88. Gülle und Tierhygiene (1975–79). "Zur Notwendigkeit des Abwasser-Gülle-Verwertungsgebietes im Raum Eberswalde." The Lichterfelde farm was a KAP, as opposed to an LPG.

31 Lutze, *Landschaft Im Wandel*, 64.

32 SAPMO BArch, DY 30/1687, pp. 187–88. Gülle und Tierhygiene (1975–79). "Zur Notwendigkeit des Abwasser-Gülle-Verwertungsgebietes im Raum Eberswalde."

33 Lutze, *Landschaft Im Wandel*, 64.

34 SAPMO BArch, DY 30/1687, pp. 187–88. Gülle und Tierhygiene (1975–79). "Zur Notwendigkeit des Abwasser-Gülle-Verwertungsgebietes im Raum Eberswalde."

35 Lutze, *Landschaft Im Wandel*, 66–67. Lutze writes in great detail about the program in his study of Brandenburg's natural history.

36 Olmstead and Rhode, *Creating Abundance*, 1–16; 262–83.

37 Shrubb, *Birds, Scythes and Combines*, 46–60.

38 Stoll, *Larding the Lean Earth*, 56.

39 Steiner, *The Plans That Failed*, 157.

40 LAB C Rep. 112 No. 282, Verwendung der Organische Düngemittel, Letter from Heinz Kuhrig to the District Council of Brandenburg, June 25, 1976.

41 LAB C Rep. 112 No. 282, Ibid.

42 LAB C Rep. 112 No. 282, ibid. "Anlage: Erste Erfahrungen bei der Produktion und dem Einsatz von Kompost," 1. The section notes, however, that all of these waste byproducts differ widely in their chemical and physical composition and as such would require significant processing until they were suitable for use.

43 LAB C Rep. 112 No. 282, 6–7. Verwendung der Organische Düngemittel, Letter from Heinz Kuhrig to the District Council of Brandenburg, June 25, 1976. "Anlage: Erste Erfahrungen bei der Produktion und dem Einsatz von Kompost."

44 BArch DK, 5/1730, pp. 29, 38. Nitrate and Water Contamination. "Studie zur Stickstoffbelastung der Gewässer der DDR, Ihre Entwicklung in den Letzten Jahren und Hauptrichtungen für Deren Veränderung," Institut für Wasserwirtschaft, Berlin, 1979.

45 LAB C Rep. 112 No. 282, p. 7. Verwendung der Organische Düngemittel, Letter from Heinz Kuhrig to the District Council of Brandenburg, June 25, 1976. "Anlage: Erste Erfahrungen bei der Produktion und dem Einsatz von Kompost."

46 Editors, "Das Aushängeschild," *Neue Deutsche Bauernzeitung* 9/1979 (March 2, 1979): 8.

47 Hildegard Sens, "Stapeldung ist Ehrensache," *Neue Deutsche Bauernzeitung* 8/1979 (February 23, 1979): 8.

48 Sens, "Stapeldung ist Ehrensache," *Neue Deutsche Bauernzeitung* 8/1979 (February 23, 1979): 8.

49 Günther Schattenberg, "Wasser Ist Gift für Gülle: Mit Wissenschaftler in der Milchviehanlage Werder, Kreis Lübz," *Neue Deutsche Bauernzeitung* (March 9, 1979): 9.

50 Bruno Loeding, "Weniger Wasser in die Gülle," *Neue Deutsche Bauernzeitung* 8/1979 (February 16, 1979): 9.

51 BArch, DK 5/1141, p. 2. Report on the "destruction and endangerment of the water supply through careless manure treatment," marked "confidential." Commissioned by the Council of Ministers on July 10, 1980 and delivered to Politburo and Council, April 8, 1981. It appears to have been read and reviewed at least every third September between 1981 and 1986.

52 BArch, DK 5/1141, p. 2. Report on the "destruction and endangerment of the water supply through careless manure treatment," marked "confidential," April 8, 1981.

53 Buechler and Buechler, *Contesting Agriculture*, 69.

54 Buechler and Buechler, *Contesting* Agriculture, 69.

55 Stoll, *Larding the Lean Earth*, 52. "At the average composition of 0.5 percent nitrogen, 0.25 percent phosphorous, and 0.5 percent potassium per unit of solid dung, it takes 2 metric tons (2.2 short tons) to equal just 100 kg of the supercharged factory-made fertilizer popular today. Farmers intent on using animal manure to duplicate the punch of synthetic mixes need eighteen to twenty tons per acre."

56 SAPMO BArch, DY 30/IV B 2/2.023/61. ABI Reports (1978–81). Report on Manure Problems, February 28, 1978, pp. 4–5.

57 Herbert Lehmann, "Güllebecken gehören dazu," *Neue Deutsche Bauernzeitung* 9/1979 (February 23, 1979): 8.

58 Fridger Pelta, "Breitgekleckert," *Neue Deutsche Bauernzeitung* 10/1979 (March 2, 1979): 8.

59 Lehmann, "Güllebecken gehören dazu," *Neue Deutsche Bauernzeitung* 9/1979 (February 23, 1979): 8.

60 Fridger Pelta, "Erst Silos und Nun Güllelager," *Neue Deutsche Bauernzeitung* 19/1982 (May 7, 1982): 6.

61 BArch DK, 5/1141, p. 3. Report on the "destruction and endangerment of the water supply through careless manure treatment," marked "confidential," April 8, 1981.

62 Fritz Carl, "Kompostieren in Hohlwegen," Berlin, Hauptstadt der DDR: *Neue Deutsche Bauernzeitung* 8/1979 (February 16, 1979): 4. In Carl's words, "Hauptsache die Behälter sind leer" (The main thing is that the containers are empty).

63 Heinz Krüger, "Keine Nachsicht bei Kahlstellen," *Neue Deutsche Bauernzeitung* 16/1979, (March 16, 1979): 4.

64 Monika Jacobs, "Versuche im Plätinsee," *Neue Deutsche Bauernzeitung* 17/1989 (April 12, 1979), 27.

65 West and Kennedy, "Swine Manure," 3–5; Edwards and Driscoll, "Environmental Consequences of Swine," 163–65.

66 SAPMO BArch, DY 30/1741, pp. 80–81. Nature and Environmental Protection, Band 2 (1980–89). "Water Pollution from Farm Runoff (1980–1982)." Study on possible solutions to farm runoff and mixture of drinking water. Institut für Düngungsforschung Leipzig-Potsdam der AdL d. DDR. August 12, 1980.

67 West and Kennedy, "Swine Manure," 3–5; Edwards and Driscoll, "Environmental Consequences of Swine," 163–65.

68 BArch DK, 5/1730, p. 1. Nitrate and Water Contamination (1979). Institute of Water Management, "Studie zur Stickstoffbelastung der Gewässer der DDR, Ihre Entwicklung in den Letzten Jahren und Hauptrichtungen für Deren Veränderung," Berlin, February 1979.

69 BArch, DK 5/1730, p. 2. Nitrate and Water Contamination (1979). Institute of Water Management, "Studie zur Stickstoffbelastung der Gewässer der DDR, Ihre Entwicklung in den Letzten Jahren und Hauptrichtungen für Deren Veränderung," Berlin, February 1979.

70 BArch, DK 5/1730, p. 61. Nitrate and Water Contamination (1979). Institute of Water Management, "Studie zur Stickstoffbelastung der Gewässer der DDR, Ihre Entwicklung in den Letzten Jahren und Hauptrichtungen für Deren Veränderung," Berlin, February 1979.

71 BArch, DK 5/1730, p. 117. Nitrate and Water Contamination (1979). Institute of Water Management, "Studie zur Stickstoffbelastung der Gewässer der DDR, Ihre Entwicklung in den Letzten Jahren und Hauptrichtungen für Deren Veränderung," Berlin, February 1979.

72 SAPMO BArch, DY 30/1741, p. 136. Nature and Environmental Protection, Band 2 (1980–89). A letter to the minister of environmental protection (Herrn Reichelt), August 11, 1980.

73 SAPMO BArch DY, 30/1741, p. 136. Nature and Environmental Protection, Band 2 (1980–89). A letter to the minister of environmental protection (Herrn Reichelt), August 11, 1980.

74 SAPMO BArch, DY 30/1741, p. 138. Nature and Environmental Protection, Band 2 (1980–89). A letter to the minister of environmental protection (Herrn Reichelt), August 11, 1980.

75 SAPMO BArch, DY 30/1741, p.138. Nature and Environmental Protection, Band 2 (1980–89). A letter to the minister of environmental protection (Herrn Reichelt), August 11, 1980.

76 Editors, "Was der Mensch so Alles Verträgt," *Der Spiegel* 17 (April 3, 1990): 43.

77 Schönfelder, *Gott Gegen Gülle*, 23–24.

78 Buechler and Buechler, *Contesting Agriculture*, 69.

79 Editors, "Was der Mensch so Alles Verträgt," *Der Spiegel* 17 (April 3, 1990): 43.

80 BArch DK, 5/1730, pp. 87–89. Nitrate and Water Contamination (1979). Institute of Water Management, "Studie zur Stickstoffbelastung der Gewässer der DDR, Ihre Entwicklung in den Letzten Jahren und Hauptrichtungen für Deren Veränderung," Berlin, February 1979.

81 BArch DK, 5/1730, 89–90. Nitrate and Water Contamination (1979). Institute of Water Management, "Studie zur Stickstoffbelastung der Gewässer der DDR, Ihre Entwicklung in den Letzten Jahren und Hauptrichtungen für Deren Veränderung," Berlin, February 1979.

82 Patricia Leigh Brown, "The Problem Is Clear: The Water Is Filthy," *New York Times*, November 13, 2012; Michael Wines, "Behind Toledo's Water Crisis, a Long-Troubled Lake Erie," *New York Times*, August 4, 2014.

83 Shiqi Yang et al., "Effects of Nitrate Leaching Caused by Swine Manure Application in the Fields of the Yellow River Irrigation Zone of Ningxia, China," *Scientific Reports*, September 12, 2017, www.nature.com/articles/s41598-017-12953-9.pdf.
84 Patel and Moore, *A History of the World in Seven Cheap Things.*
85 BArch DK, 5/1730, p. 10. Nitrate and Water Contamination (1979). Institute of Water Management, "Studie zur Stickstoffbelastung der Gewässer der DDR, Ihre Entwicklung in den Letzten Jahren und Hauptrichtungen für Deren Veränderung," Berlin, February 1979.

CHAPTER FIVE: PIGS IN THE SMALL GARDEN PARADISE

1 Dietrich, *Hammer, Zirkel, Gartenzaun*, 249–52. I have drawn several primary sources for this chapter from Dietrich's voluminous annotated collection on the history and politics of gardening in the GDR. This story, for example, is retold by Dietrich, who also provides a reproduction of the report, written by G. Trölitysch, Central Committee member in charge of construction.
2 Dietrich, *Hammer, Zirkel, Gartenzaun*, 249–52. They were Heinz Kuhrig, the minister of agriculture; Comrade Schmolinksy, the regional minister of agriculture in Frankfurt am Oder; and Gerhard Grüneberg, the party secretary of agriculture.
3 SAPMO BArch, DY 30/2841; SAPMO BArch DY 30/1859, p. 200, in Dietrich, *Hammer, Zirkel, Gartenzaun*, 249–52.
4 SAPMO BArch, DY 30/IV B 2/2.023/35, p. 3. Meat Supply and Foot and Mouth Disease Reports (1975). Survey produced by Bezirksleitung for Neubrandenburg, Potsdam, Rostock, Schwerin, "Information über Meinungen zu den neuen ökonomischen Regelungen in der Land- und Nahrungsgüterwirtschaft," December 11, 1975.
5 SAPMO BArch, DY 30/1543, pp. 130–31. Individual Livestock and Produce Production (1982–1989). Klaus Siegemund, *Die Individuelle Produktion Landwirtschaftlicher Erzeugnisse.*
6 Dietrich, *Hammer, Zirkel, Gartenzaun*, 13.
7 Schier, *Alltagsleben im Sozialistischen Dorf*, 230–31.
8 SAPMO BArch, DY 30/1860. VKSK, Band 8 (1980–89); SAPMO BArch, DY 14, introduction to finding aid; Dietrich, *Hammer Zirkel, Gartenzaun*, 13; Schier, *Alltagsleben im "Sozialistischen Dorf,"* 225.
9 BArch, DY 30/1860. VKSK, Band 8 (1980–89); SAPMO BArch, DY 14, Introduction to finding aid.
10 Mary Fulbrook, *The People's State*, 57.
11 Dietrich, *Hammer, Zirkel, Gartenzaun*, 13; Schier, *Alltagsleben im "Sozialistischen Dorf,"* 225.
12 The GDR was not the only communist country to have a robust garden culture in the middle of the planned economy. On Cuba, see Premat, *Sowing Change.* Postwar France saw an analogous embrace of small-scale organic farming in

the wake of the oil crises of the 1970s, as detailed by Venus Bivar in *Organic Resistance*, 141–69.

13 SAPMO BArch, DY 14/90. Nachlass Wilhelm Groh, Band 1 (1920–78). Sigurd Darac, "Wandlung einer Laubenkolonie," *Die Wochenpost*, undated, 16–17.

14 SAPMO BArch, DY 14/90. Nachlass Wilhelm Groh, Band 1 (1920–78). Sigurd Darac, "Wandlung einer Laubenkolonie," *Die Wochenpost*, undated, 16–17.

15 And like the industrial revolution itself, the first pauper gardens began in England.

16 SAPMO BArch, DY 14/90. Nachlass Wilhelm Groh, Band 1 (1920–78). Johannes Gerta Nuhs, "Daniel Schreber—Geistiger Vater der Kleingärten," *Neue Deutsche Bauernzeitung* 47/1981 (November 20, 1981): 12–13.

17 SAPMO BArch, DY 14/90. Nachlass Wilhelm Groh, Band 1 (1920–78). Johannes Gerta Nuhs, "Daniel Schreber—Geistiger Vater der Kleingärten, *Neue Deutsche Bauernzeitung* 47/1981 (November 20, 1981): 19.

18 Haney, *When Modern Was Green*, 99–104.

19 Haney, *When Modern Was Green*, 113.

20 Leberecht Migge, *Jedermann Selbstversorger: Eine Lösung der Siedlungsfrage durch Neuen Gartenbau*, (Jena: Diedrichs, 1918) 1, in Haney, *When Modern Was Green*, 114–17. *Everyman Self-Sufficient*! also drew on the work of the Russian prince and anarchist Peter Kropotkin.

21 Leberecht Migge, "Die Grüne Manifest," *Die Tat* 10, no. 2 (1918–19): 912–19, in Haney, *When Modern Was Green*, 122–23.

22 Haney, *When Modern Was Green*, 125–26.

23 Harsch, "Versorgungspolitik in der Sowjetisch Besetzten Zone."

24 Fulbrook, *The People's State*, 72.

25 Fulbrook, *The People's State*, 62.

26 BArch, DY 14/28, p. 11. Beschlussvorlagen für Sitzungen des Präsidiums und des Sekretariats zur Auswertung Wettbewerbsergebnisse und zur Auswertung der Grundkartei, Band 1 (1971–85). Annual Yields (1971–1984), December 31, 1984.

27 Buechler and Buechler, *Contesting Agriculture*, 84; Schier, *Alltagsleben Im "Sozialistischen Dorf*," 70–71.

28 Buechler and Buechler, *Contesting Agriculture*, 45–46.

29 Schier, *Alltagsleben Im "Sozialistischen Dorf*," 223–24, 230–331.

30 Buechler and Buechler, *Contesting Agriculture*, 85.

31 Rolf Urland, "Fleisch für acht Menschen aus der Stadt," *Neue Deutsche Bauernzeitung* 51/1981 (December 18, 1981) 6.

32 Urland, "Fleisch für acht Menschen aus der Stadt," 6.

33 Urland, "Fleisch für acht Menschen aus der Stadt," 6.

34 Urland, "Fleisch für acht Menschen aus der Stadt," 6.

35 See "Ulbricht's Veto" in Dietrich, *Hammer, Zirkel, Gartenzaun*, 76–78.

36 Lebow, *Unfinished Utopia*, 33.

37 Harsch, *Revenge of the Domestic*, 167–68.

38 SAPMO BArch, DY 14. Introduction to finding aid.

39 SAPMO BArch, DY 14. Introduction to finding aid.

40 The German word here, *spießig*, refers to bourgeois culture. The best approximation in English might be the vernacular "bougie," or "stuffy."

41 Marlis Allendorf, "Ich hab 'ne Laube, bin ich Spiesser?" (*FÜR DICH*, 17/1975, 38), in Dietrich, *Hammer, Zirkel, Gartenzaun*, 208.

42 SAPMO Barch, DY 30/IV/B 2/2/023/35. Fleisch Versorgung (1975). "Konzeption für die Entwicklung der Landwirtschaftlichen Produktion und der Versorgung der Hauptstadt der DDR, Berlin, 1975."

43 SAPMO BArch, DY 30/1857, p. 32, in Dietrich, *Hammer, Zirkel, Gartenzaun*, 208.

44 LAB C Rep. 112/252, Analysis of Petitions for Food Production and Agriculture (1975–1980). Magistrate of Greater Berlin, Department of Agriculture and Food to the first representative of the Oberbürgermeister (Comrade Horst Palm), 2–3.

45 SAPMO BArch, DY 30/J IV 2/3/2476, p. 1, in Dietrich, *Hammer, Zirkel, Gartenzaun*, 230.

46 *Neues Deutschland* 14 (September 1976): 3, in Dietrich, *Hammer, Zirkel, Gartenzaun*, 235.

47 For detailed discussion of this program, see chapter 2.

48 SAPMO BArch, DY 30/1858, p. 112. VKSK, Band 6, January–July, 1977, in Dietrich, *Hammer, Zirkel, Gartenzaun*, 236–37.

49 Jenny Leigh Smith, "Empire of Ice Cream: How Life Got Sweet in the Postwar Soviet Union," in Belasco and Horowitz, *Food Chains*, 142–59.

50 Egon Seidel und Helga Schulte, "Welche Bedeutung hat der Aufbau eines Gemüsegürtels zur Versorgung eines Verbraucherzentrums," *Kooperation 8* (Berlin, Hauptstadt der DDR: Council for Agricultural Production and Food, vol. 3, 1974), 119.

51 SAPMO BArch, DY 30/IV/ B 2/2.023/35. Meat Supply and Foot and Mouth Disease Reports (1975). Letter from Kiesler to Grüneberg, July 25, 1975; "Konzeption für die Entwicklung der Land-wirtschaftlichen Produktion und der Versorgung der Hauptstadt der DDR, Berlin," July 1975, 5.

52 Seidel and Schutel, "Welche Bedeutung hat der Aufbau," 120–21.

53 SAPMO BArch, DY 30/IV/ B 2/2.023/35, p. 125. Meat Supply and Foot and Mouth Disease Reports (1975). Letter from Kiesler to Grüneberg, July 25, 1975.

54 SAPMO BArch, DY 30/IV/ B 2/2.023/35, p. 125. Meat Supply and Foot and Mouth Disease Reports (1975). Letter from Kiesler to Grüneberg, July 25, 1975.

55 Schier, *Alltagsleben Im "Sozialistischen Dorf,"* 225.

56 SAPMO BArch, DY 30/16739. ABI (Arbeiter- und Bauerninspektion), Band 4 (1979–81). "Information over the Results of an Inspection of the Further Improvement of the Supply of Fresh Vegetables," Committee of the ABI, October 28, 1981, 5.

57 SAPMO BArch, DY 30/16739. ABI, Band 4 (1979–81). "Information over the Results of an Inspection of the Further Improvement of the Supply of Fresh Vegetables," Committee of the ABI, October 28, 1981, 5.

58 SAPMO BArch DY 30/16739. ABI, Band 4 (1979–81). "Information over the Results of an Inspection of the Further Improvement of the Supply of Fresh Vegetables," Committee of the ABI, October 28, 1981, 5.

59 Protokoll des X. Parteitages der Sozialistischen Einheitspartei Deutschlands, Berlin, 1981, 1:56, in Dietrich, *Hammer, Zirkel, Gartenzaun*, 261.

60 BArch, DC 20-I/3/1916, p. 137. Resolutions and Meeting Reports. "Sitzung des Ministerrat vom 24. February 1983," vol. 2, *Reduction of Produce Transportation Costs 1982*.

61 BArch DC, 20-I/3/1902, p. 42. Resolutions and Meeting Reports, Anlage 4a. "Sitzung des Ministerrat vom 23. Dezember 1982," vol. 2, *Reduction of Produce Transportation Costs, 1982*.

62 BArch DC, 20-I/3/1902, p. 42. Resolutions and Meeting Reports, Anlage 4a. "Sitzung des Ministerrat vom 23. Dezember 1982," vol. 2, *Reduction of Produce Transportation Costs, 1982*.

63 BArch DC, 20-I/3/1916, p. 137. Eigenversorgung mit Gemüse, Berlin (1983). "Beschluß über Maßnahmen zur Sicherung der Geplanten Gemüseproduktion 1983 und zur Weiteren Erhöhung der Eigenversorgung mit Gemüse in Bezirken, Kreisen, und Gemeinden vom 24. Februar 1983," Anlage 2, p.1.

64 BArch DC 20-I/3/1916, pp. 144–45. Eigenversorgung mit Gemüse, Berlin (1983). "Beschluß über Maßnahmen zur Sicherung der Geplanten Gemüseproduktion 1983 und zur Weiteren Erhöhung der Eigenversorgung mit Gemüse in Bezirken, Kreisen, und Gemeinden vom 24. Februar 1983," Anlage 2,

65 BArch DC 20-I/3/1916, p. 173.

66 Fridger Pelta, "Die Neue Sparte übernimmt den Hang," *Neue Deutsche Bauernzeitung* 40/1981 (October 2, 1981): 10.

67 Fridger Pelta, "Die Neue Sparte übernimmt den Hang," *Neue Deutsche Bauernzeitung* 40/1981 (October 2, 1981): 10.

68 Margitta Görner, "Ein Mietenplatz wird Küchengarten," *Neue Deutsche Bauernzeitung* 31/1979 (August 3, 1979): 6.

69 Wolfgang Warzock, "Zwei Prämienläufer von der Genossenschaft," *Neue Deutsche Bauernzeitung* 31/1979 (August 3, 1979): 11.

70 SAPMO BArch, DY 30/1860, pp. 21–22. VKSK Eingaben, Band 8 (1980–89). VKSK letter to Erich Honecker, November 6, 1980.

71 SAPMO BArch, DY 30/1860, p. 33. VKSK Eingaben, Band 8 (1980–89). "Staatliches Aufkommen der Kleinproduzenten," April 7, 1981.

72 SAPMO BArch DY, 30/1860, p. 21. VKSK Eingaben, Band 8 (1980–89). Letter to Erich Honecker, November 6, 1980.

73 SAPMO BArch, DY 30/J IV 2/2/1973, p. 31. "Beschluß über die Durchführung der Agrarpreisreform," in Dietrich, *Hammer Zirkel*, 282.

74 Dietrich, *Hammer, Zirkel, Gartenzaun*, 283.

75 Dietrich, *Hammer, Zirkel, Gartenzaun*, 288.

76 SAPMO BArch, DY 30/1524, p. 227. Ministerium für Handel und Versorgung. "Information über die Verfütterung von Nahrungsgütern und die Sich daraus Ergebenden Folgewirkungen für den Konsumgüterbinnenhandel bei der

Durchführung der Agrarpreisreform," March 1983, in Dietrich, *Hammer, Zirkel, Gartenzaun*, 289–90.

77 Traditionally, the winter months served as peak slaughtering season.

78 SAPMO BArch DY, 30/1543, pp. 7–8. Individual Livestock and Produce Production (1982–89). Fourth Quarter Report, October 19, 1983.

79 SAPMO BArch DY, 30/1543, pp. 25–28. Individual Livestock and Produce Production (1982–89). First Quarter Report, 1984. February 3, 1984.

80 SAPMO BArch DY 30/1543, pp. 25–28. Individual Livestock and Produce Production (1982–89). First Quarter Report, 1984. February 3, 1984. The decrease in household slaughtering fell by 51,000 to 206,000 in 1984.

81 Schier, *Alltagsleben im "Sozialistischen Dorf,"* 230.

CHAPTER SIX: A PLAGUE OF WILD BOARS

1 Portions of this chapter are reprinted from Fleischman, "'A Plague of Wild Boars,'" 1015–34.

2 LAB C Rep. 112 No. 240. Magistrat von Berlin, Abteilung Land-, Forst- und Nahrungsgüterwirtschaft. Wildschwein Plage (1978–89). Helmut Arndt, Berlin, writing to the minister of agriculture and forestry, July 17, 1988.

3 LAB C Rep. 112 No. 240. Magistrat von Berlin, Abteilung Land-, Forst- und Nahrungsgüterwirtschaft. Wildschwein Plage (1978–89). Helmut Arndt, Berlin, writing to the minister of agriculture and forestry, July 17, 1988

4 LAB C Rep. 112 No. 240. Magistrat von Berlin, Abteilung Land-, Forst- und Nahrungsgüterwirtschaft. Wildschwein Plage (1978–89). Helmut Arndt, Berlin, writing to the minister of agriculture and forestry, July 17, 1988. As shown by the name of this file, "Wildschwein Plage."

5 Meynhardt, *Schwarzwild Report*.

6 Stubbe, "Wildbewirtschaftung," 69.

7 Stubbe, "Wildbewirtschaftung," 70.

8 Meynhardt, *Schwarzwild Report*, 188.

9 Melville, *A Plague of Sheep*, 6–7.

10 Moranda, *The People's Own Landscape*.

11 Stubbe, "Wildbewirtschaftung," 75.

12 Crosby, *Ecological Imperialism*, 173; Walker, "Commercial Growth and Environmental Change."

13 Meynhardt, *Schwarzwild Report*, 14.

14 Meynhardt, *Schwarzwild Report*, 29.

15 Meynhardt, *Schwarzwild Report*, 167.

16 Ciesla and Suter, *Jagd und Macht*, 198–202; Filip Slaveski, *The Soviet Occupation of Germany*; Naimark, *The Russians in Germany*, 198–202.

17 Scott, *Seeing like a State*, 11–22.

18 Nelson, *Cold War Ecology*, 23.

19 Schabel, "Deer and Dauerwald in Germany"; Herf, *Reactionary Modernism;* Haney, *When Modern Was Green*.

20 Nelson, *Cold War Ecology*, 19–24.
21 Uekötter, *The Green and the Brown*; Bramwell, *Blood and Soil.*
22 Stubbe, "Wildbewirtschaftung," 69.
23 Stubbe, "Historischer Abriss Der Organisation Des Jagdwesens," 20–25.
24 Moranda, *The People's Own.*
25 Lekan, *Imagining the Nation in Nature.*
26 Moranda, *The People's Own*, 63–64.
27 Stubbe, "Historischer Abriss der Organisation des Jagdwesens," 28.
28 Stubbe, "Ausbildung und Schulung," 281.
29 Meynhardt, *Schwarzwild Report*, 162.
30 Nelson, *Cold War Ecology*, 74–76.
31 Ciesla and Suter, *Jagd und Macht*, 137–90.
32 Fulbrook, *The People's State*, 81.
33 Trömer, "Gesetzgebung, Organisation und Leitung des Jagdwesens Ab 1962," 32–33, 48.
34 Haselmann, "Die Jagd in der DDR," 40–41.
35 Haselmann, "Die Jagd in der DDR," 41–42.
36 Ciesla and Suter, *Jagd und Macht*, 206–7.
37 Ciesla and Suter, *Jagd und Macht*, 206–7.
38 Ciesla and Suter, *Jagd und Macht*, 217–23.
39 BArch, DK 1/26682. Handwritten note, author unknown. Ministry of Agriculture, Forestry, and Food. Official Program for the Rabbit Hunt in Bezirk Magdeburg, December 7, 1985. This was highlighted and presented in a special online publication for the German Federal Archives by archivists Silke Liberona-Chamorro and Heike Zeise, titled "Hasenjagd der Diplomaten." www.bundesarchiv.de/DE/Content/Virtuelle-Ausstellungen/Hasenjagd-Der-Diplomaten/hasenjagd-der-diplomaten.html.
40 Ciesla and Suter, *Jagd und Macht*, 226–27.
41 Fulbrook, *The People's State*, 226–27.
42 Moranda, *The People's Own*, 159–60.
43 Stubbe, "Wildbewirtschaftung," 94.
44 James B. Campbell, *Introduction to Remote Sensing.*
45 Schabel, "Deer and Dauerwald in Germany," 888.
46 Lutz Briedermann, *Der Wildbestand, die Grosse Unbekannte*, 1.
47 Stubbe, "Wildbewirtschaftung," 93–94.
48 BArch, DK 107/33347. Scientific Studies of Wild Boar Rearing. "Erarbeitung wissenschaftlicher Grundlagen zur Sicherung einer Nachhaltigen Produktivität in Schwarzwild-Bewirtschaftungsgebieten, Abschlußbericht von Dr. Briedermann, L.," 1980, 10.
49 Meynhardt, *Schwarzwild Report*, 44.
50 BArch, DK 107/33347. Scientific Studies of Wild Boar Rearing. "Erarbeitung wissenschaftlicher Grundlagen zur Sicherung einer Nachhaltigen Produktivität in Schwarzwild-Bewirtschaftungsgebieten, Abschlußbericht von Dr. Briedermann, L.," 1980, 10.

51 Virgos, "Factors Affecting Wild Boar (Sus Scrofa) Occurrence."
52 SAPMO-BArch, DY 30/1789, p. 40. Forestry and Hunting Matters, Band 11 (1981–83). Letter from the LPG P "Saletal" Sitz Kahla, July 29, 1981.
53 Stubbe, "Wildbewirtschaftung," 113–14.
54 Meng, Lindsay, and Sriranganathan, "Wild Boars as Sources for Infectious Diseases."
55 Francis, *Domesticated.*
56 White, "From Globalized Pig Breeds to Capitalist Pigs"; Mizelle, *Pig*, 26–40.
57 White, "Globalized Pig Breeds," 98–100; Meynhardt, *Schwarzwild Report*, 17.
58 Watson, *The Whole Hog*, 108.
59 Meynhardt, *Schwarzwild Report*, 17.
60 Mayer and Brisbin, *Wild Pigs in the United States*, 124–40, 190, 225–27.
61 Frantz, Massei, and Burke, "Genetic Evidence for Past Hybridization."
62 Meynhardt, *Schwarzwild Report*, 46.
63 LAB C Rep. 112 No. 340. "Maßnahmen zur Bejagung von Schwarzwild auf dem Territorium der Haupstadt der DDR, Berlin." December 4, 1984. The order stated "Für das Territiorium der Haupstadt und Deren Jagdgebiete sind Keine Bejagung- und Bewirtschaftsrichtlinien Zuverlässig."
64 Stubbe, "Wildbewirtschaftung," 99–100.

CHAPTER SEVEN: THE IRON LAW OF EXPORTS

1 Editor, "Mildes Wetter zum Jahresbeginn," *Berliner Zeitung* no. 1 (January 2/3, 1982): 1.
2 SAPMO-BArch, DY 30/1689, pp. 1–2. Monthly, Quarterly, and Annual Results in Livestock Production, Band 1 (1982). Report from January 4, 1982.
3 "Transmissible Gastroenteritis in Swine," Indiana Animal Disease Diagnostic Laboratory, Purdue University, fall 2008 newsletter, www.addl.purdue.edu /newsletters /2008/Fall/TGE.htm.
4 SAPMO-BArch, DY 30/1696, pp. 16–18. Veterinary Matters, Epizootics. Werner Linder, Leiter Veterinärwesen, writes to Felfe, January 19, 1982.
5 Billy Flowers, "Take a Closer Look at Total Born, Mummies and Stillborns," *National Hog Farmer*, last updated February 27, 2017, www.nationalhogfarmer .com/animal-health/take-closer-look-total-born-mummies-and-stillborns.
6 "Classical Swine Fever or Hog Cholera," *Pig Site*, accessed on May 12, 2013, www.thepigsite.com/pighealth/article/447/classical-swine-fever-csf-hog -cholera-hc.
7 SAPMO-BArch, DY 30/1696, pp. 19–20. Veterinary Matters, Epizootics. Report, "Schweinepest in der LPG Gnoien, Kreis Teterow, 14.1.1982."
8 SAPMO-BArch, DY 30/1689, pp. 39–42. Monthly, Quarterly, and Annual Results in Livestock Production, Band 1 (1982). Report addressed to Felfe, January 22, 1982.
9 SAPMO-BArch, DY 30/1689, pp. 99–100. Monthly, Quarterly, and Annual Results in Livestock Production, Band 1 (1982). January Production Numbers,

February 18, 1982. Sows losses in January rose by 163 percent of the previous year's numbers to 14,501 animals.

10 BArch, DC 20/20813, p. 38. Staatssekretär/Arbeitsgruppe für Organisation und Inspektion (1972–89). Berichte der Kontrollabteilung. Außenhandel, insb. NSW-Export und Import. Willi Stoph Speech, "In der Beratung mit den Vorsitzenden der Räte der Bezirke, am 28.4.1982."

11 Kopstein, *The Politics of Economic Decline in East Germany*, 92.

12 Kopstein, *The Politics of Economic Decline in East Germany*, 92–94.

13 Zatlin, *The Currency of Socialism*, 139.

14 Dale, *Between State Capitalism and Globalisation*, 200.

15 Schmidt and Stern, *Unser Jahrhundert*, 183; Editors, "Ein Kanzler zu Besuch beim Staatsratsvorsitzenden," Mitteldeutsche Rundfunk, www.mdr.de/damals/archiv/schmidt-in-der-ddr102.html.

16 Judt, *Postwar*, 586.

17 Zatlin, *The Currency of Socialism*, 125–28.

18 SAPMO-BArch, DY 30/J IV 2/2A/2477, p. 170. Sitzung des Politbüros, Protokoll No. 21/82, 25. Mai, 1982. Grain imports were 3.5 million tons in 1981, were planned to fall to 2.5 million tons in 1982, and were planned to fall again to 1.6 million tons in 1983.

19 SAPMO-BArch, DY 30/1689, pp. 1–2. Monthly, Quarterly, and Annual Results in Livestock Production, Band 1 (1982). Report to Felfe, January 4, 1982.

20 SAPMO-BArch, DY 30/1689, pp. 161–63. Monthly, Quarterly, and Annual Results in Livestock Production, Band 1 (1982). Report to Felfe, March 10, 1982.

21 SAPMO-BArch, DY 30/1689, p. 82. Monthly, Quarterly, and Annual Results in Livestock Production, Band 1 (1982). "Veterinary Report," February 5, 1982; BArch, DC 20/10731. Report, "Information zur Sicherung der Versorgung der Bevölkerung mit wichtigen Nahrungsmitteln," March 29, 1982. Attachment, assessing regional situation, including LPG reports, p. 4.

22 SAPMO-BArch, DY 30/1689, p. 74. Monthly, Quarterly, and Annual Results in Livestock Production, Band 1 (1982). "Veterinary Report," February 2, 1982.

23 The conceptualization was not unique to East Germany as Blanchette has shown. Blanchette, "Herding Species," 640–69.

24 SAPMO-BArch, DY 30/1689, pp. 99–100. Monthly, Quarterly, and Annual Results in Livestock Production, Band 1 (1982). Report, January Production Numbers, February 18, 1982.

25 SAPMO-BArch, DY 30/1689, p. 82. Monthly, Quarterly, and Annual Results in Livestock Production, Band 1 (1982). "Veterinary Report," February 5, 1982.

26 SAPMO-BArch, DY 30/1689, p. 39. Monthly, Quarterly, and Annual Results in Livestock Production, Band 1 (1982). "Veterinary Report," January 22, 1982.

27 SAPMO BArch, DY 30/1689, p. 139. Monthly, Quarterly, and Annual Results in Livestock Production, Band 1 (1982). "Veterinary Report," February 1982.

28 BArch DC, 20/10730, p. 4. Working Group for Organization and Inspection. Report, "Information about Livestock Losses of Cattle and Pigs, as of February 28, 1982." Produced in March of 1982; SAPMO-BArch, DY 30/1689, p. 75.

Monthly, Quarterly, and Annual Results in Livestock Production, Band 1 (1982). Veterinary Report, February 5, 1982. The Ministry of Agriculture also noted that as of February 5, Halle had seven hundred animals waiting to be picked up.

29 SAPMO-BArch, DY 30/1533, p. 23. "Analyses of the Provisioning of Food," KLW VVB, Report on Growing Shortages, February 3, 1982.

30 Woods, *A Manufactured Plague?*, xiii. Symptoms drawn from Wood's description of FMD in her book.

31 SAPMO-BArch, DY 30/1696, p. 79. Veterinary Matters, Epizootics. "Veterinary Report on the Outbreak of Foot and Mouth Disease," March 15, 1982. In German, the disease is known as *Maul und Klauenseuche*, or MKS.

32 SAPMO-BArch, DY 30/1696, p. 79. Veterinary Matters, Epizootics. "Veterinary Report on the Outbreak of Foot and Mouth Disease," March 15, 1982.

33 SAPMO-BArch DY, 30/1696, p. 79. Veterinary Matters, Epizootics. "Veterinary Report on the Outbreak of Foot and Mouth Disease," March 15, 1982.

34 BArch DC, 20/10731. Reports and Investigations. Report, "Information zur Entwicklung und Wirksamkeit der Bekämpfung von Tierseuchen 1981," April 2, 1982, 6–7.

35 LAB C Rep. 112 No. 345, pp. 1–2. Magistrat von Berlin, Abteilung Land-, Forst- und Nahrungsguterwirtschaft. Anweisung No. 2/82, March 9, 1982.

36 SAPMO-BArch, DY 30/1696, p. 84. Report, Heinz Kuhrig to Bruno Lietz, March 17, 1982.

37 SAPMO-BArch, DY 30/1696, pp. 73–74. Report, Heinz Kuhrig to Bruno Lietz, March 18, 1982.

38 SAPMO-BArch DY 30/1696, pp. 73–74. Report, Heinz Kuhrig to Bruno Lietz, March 18, 1982.

39 Woods, *Manufactured Plague*.

40 Woods, *Manufactured Plague*, 71.

41 SAPMO-BArch, DY 30/1696, pp. 225–26. Report from Soviet veterinarians, April 20, 1982.

42 Martina Rathke, "Forschungsinstitut Vermütlich an MKS-Ausbruch Schuld," *Hamburger Abendblatt*, September 20, 2010. www.abendblatt.de/region/norddeutschland/article1637406/Forschungsinstitut-vermutlich-an-MKS-Ausbruch-schuld.html.

43 Rathke, "MKS-Ausbruch."

44 SAPMO-BArch, DY 30/1696, pp. 225–26. Report from Soviet veterinarians, April 20, 1982. They were the Soviet institute's director (one Comrade Dudnikow), an assistant researcher (one Comrade Schaschkow), and the Soviet Union's chief veterinarian from the Ministry of Agriculture, (one Comrade Tretjakow).

45 Rathke, "MKS-Ausbruch."

46 SAPMO-BArch, DY 30/1696, pp. 92–93. Veterinary Matters, Epizootics. "Report for Erich Honecker, March 22, 1982."

47 SAPMO-BArch, DY, 30/1696, pp. 92–93. Veterinary Matters, Epizootics. "Report for Erich Honecker, March 22, 1982." The rest of the Hungarian regime remained less sympathetic, agreeing only to send a veterinary specialist to a consultation in the GDR on March 23.

48 SAPMO-BArch, DY, 30/1696, p. 225. Report from Soviet veterinarians, April 20, 1982.

49 SAPMO-BArch, DY 30/IV B 2/2.023/35. Fleisch Versorgung and FAM Disease Reports (1975-). Letter from the "Oberbürgermeister" of Berlin to Genosssen Konrad Naumann, Kandidat des Politbüros und 1. Sekretär der Bezirksleitung der SED, March 6, 1975.

50 SAPMO-BArch DY, 30/IV B 2/2/023/36. FMD outbreaks and Land Pressures on Cultivation Patterns (1975–76). Heinz Kuhrig to Grüneberg, April 23, 1976.

51 Schürer, *Gewagt und Verloren*, 141–42.

52 BArch DC, 20/20813, p. 38. Staatssekretär/Arbeitsgruppe für Organisation und Inspektion (1972-1989). Berichte der Kontrollabteilung. Außenhandel, insb. NSW-Export und Import. Willi Stoph Speech, "In der Beratung mit den Vorsitzenden der Räte der Bezirke, am 28.4.1982."

53 Beaumont, *Die DDR*, 124.

54 Zatlin, *The Currency of Socialism*, 136–39.

55 Zatlin, *The Currency of Socialism*, 136–39.

56 BArch DC, 20/10731, Council of Ministers. Reports and Investigations. Transcript from Ministerrat, March 25, 1982.

57 SAPMO-BArch, DY 30/J IV 2/2A/2465, p. 59. Protokoll No. 13/82. Sitzung des Politbüros am 30. März 1982. Arbeitsprotokoll. Kuhrig's presentation.

58 SAPMO-BArch, DY 30/J IV 2/2A/2465, p. 59. Protokoll No. 13/82. Sitzung des Politbüros am 30. März 1982. Arbeitsprotokoll. Kuhrig's presentation.

59 SAPMO-BArch, DY 30/J IV 2/2A/2465, p. 60. Protokoll No. 13/82. Sitzung des Politbüros am 30. März 1982. Arbeitsprotokoll. Kuhrig's presentation.

60 SAPMO-BArch, DY 30/J IV 2/2A/2465, p. 65. Protokoll No. 13/82. Sitzung des Politbüros am 30. März 1982. Arbeitsprotokoll. Kuhrig's presentation.

61 SAPMO-BArch, DY 30/J IV 2/2A/2465, p. 37. Protokoll No. 13/82. Sitzung des Politbüros am 30. März 1982. Arbeitsprotokoll. Schürer presentation.

62 BArch DC, 20/20813, p. 38. Staatssekretär/Arbeitsgruppe für Organisation und Inspektion. Berichte der Kontrollabteilung. Außenhandel, insb. NSW-Export und Import. Willi Stoph Speech, "In der Beratung mit den Vorsitzenden der Räte der Bezirke, am 28.4.1982."

63 SAPMO-BArch, DY 30/1689, p. 151. Monthly, Quarterly, and Annual Results in Livestock Production, Band 1 (1982). Report, "Planerfüllung Ergebnisse für die Abteilung Landwirtschaft des ZK, 3. März, 1982."

64 SAPMO-BArch, DY 30/1689, p. 190. Monthly, Quarterly, and Annual Results in Livestock Production, Band 1 (1982). Economic Plan update, April 8, 1982.

65 BArch, DC 20/20813, pp. 40–41. Staatssekretär/Arbeitsgruppe für Organisation und Inspektion Berichte der Kontrollabteilung. Außenhandel, insb.

NSW-Export und Import. Export goes before Self-Provisioning is translated from "*Ausfuhr geht vor Eigenversorung*."

66 SAPMO-BArch DY, 30/1689, p. 201. Monthly, Quarterly, and Annual Results in Livestock Production, Band 1 (1982). Meeting description, Werner Felfe with Genosse Ostmann, April 30, 1982.

67 BArch, DC 20/20812. "Report over the Results of the Inspection of the Transportation of Living Livestock and Meat," February 2, 1982, p. 4. The GDR's most important slaughterhouses were Perleberg, Teterow, Berlin, SVK Eberswlade, and Torgau.

68 SAPMO-BArch, DY 30/1696, p. 96. Veterinary Matters, Epizootics. "Maßnahmen zur Sicherung der Versorgung unter den Gegenwärtigen Seuchebedingungen: Berlin, den 24.3.1982."

69 SAPMO-BArch DY 30/1696, 96. Veterinary Matters, Epizootics. "Maßnahmen zur Sicherung der Versorgung unter den Gegenwärtigen Seuchebedingungen: Berlin, den 24.3.1982."

70 BArch, DC 20/10731. Transcript of Ministerrat meeting, March 25, 1982, p. 13.

71 LAB Berlin, C Rep. 902 No. 5456. Versorgung in Berlin 1982. "Zur Fleischversorgung Möchte Ich Noch auf Folgende Probleme Hinweisen,"April 30, 1982, 5.

72 LAB Berlin, C Rep. 902 No. 5456. Versorgung in Berlin 1982. "Zur Fleischversorgung Möchte Ich Noch auf Folgende Probleme Hinweisen,"April 30, 1982, 5.

73 LAB C Rep. 902 No. 5456. Versorgung in Berlin 1982. "Fernmündliche Information des Abteilungsleiters im Ministerium für Handel und Versorgung, Gen. Zacher, am Freitage, April 30, 1982.

74 LAB Berlin, C Rep. 902 No. 5456. Versorgung in Berlin 1982. "Zur Fleischversorgung Möchte Ich Noch auf Folgende Probleme Hinweisen," April 30, 1982, 5.

75 SAPMO-BArch, DY 30/1533, p. 209. A discussion with party members and LPG leaders in Schwerin of the Supply Problems, September 9, 1982. The reference to "occasional customers" is translated from the German, *Laufkundschaft*.

76 SAPMO-BArch, DY 30/1533, p. 209. A discussion with party members and LPG leaders in Schwerin of the supply problems, September 9, 1982.

77 Harsch, *Revenge of the Domestic*.

78 SAPMO-BArch, DY 30/1533, p. 98. Bruno Lietz writing to Kuhrig on May 5, 1982.

79 SAPMO-BArch, DY 30/18119. "Supply Petitions from the City of Leipzig," Eingaben, Heinz-Ralf S., April 29, 1982.

80 SAPMO-BArch, DY 30/1533, pp. 99–100. Bruno Lietz writing to Kuhrig, providing excerpts from regional leaders on the situations in their parts of the country, May 5, 1982.

81 SAPMO-BArch, DY 30/1533, pp. 100–1. Bruno Lietz writing to Kuhrig, providing excerpts from regional leaders on the situations in their parts of the country, May 5, 1982.

82 SAPMO-BArch, DY 30/1533, p. 102. Bruno Lietz writing to Kuhrig, providing excerpts from regional leaders on the situations in their parts of the country, May 5, 1982.

83 SAPMO-BArch, DY 30/1533, pp. 100–1. Bruno Lietz writing to Kuhrig, providing excerpts from regional leaders on the situations in their parts of the country, May 5, 1982.

84 SAPMO-BArch, DY 30/1533, p. 92. "Bericht über einen Kontrolleinsatz am 6. Mai 1982 in Bezirk Schwerin," May 7, 1982.

85 SAPMO-BArch, DY 30/1689, p. 231, 267. Monthly, Quarterly, and Annual Results in Livestock Production, Band 1 (1982). Two reports, one from May 25, 1982, shows stabilization in pig population, and another, from July 8, claimed June was first month the plan had been over-fulfilled.

86 SAPMO-BArch, DY 30/1689, p. 337. Monthly, Quarterly, and Annual Results in Livestock Production, Band 1 (1982). Report, Kuhrig to Felfe, "Bericht über den Stand der Verwirklichung des Beschlusses zur Radikalen Senkung der Tierverluste," August 31, 1982.

87 SAPMO-BArch, DY 30/1689, p. 383. Monthly, Quarterly, and Annual Results in Livestock Production, Band 1 (1982). Kuhrig to Felfe, Report from October 22, 1982.

88 SAPMO-BArch, DY 30/1533, pp. 133–34. Food Production and Supply (1982). Heinz Kuhrig (MLFN) writing to Gen. Schürer (SPK), May 3, 1982.

89 SAPMO-BArch, DY 30/1533, pp. 73-74. Food Production and Supply (1982). "Maßnahmen zur Verwirklichung des Beschlusses des Politbüros des ZK der SED vom 6.4.1982, zur Weiteren Durchführung des Volkswirtschaftsplanes in der Tierproduktion und zur Sicherung der Versorgung mit Fleisch und Butter."

90 SAPMO-BArch DY, 30/1533, p. 149. Food Production and Supply (1982). Letter to Willi Stoph, chairman of the Council of Ministers, June 25, 1982.

91 SAPMO-BArch, DY 30/1533, p. 263. Food Production and Supply (1982). Heinz Kuhrig reporting to Werner Felfe, October 21, 1982.

92 SAPMO-BArch, DY 30/1533, pp. 138–40. Food Production and Supply (1982). "Handmaterial für Genossen Felfe zur Beratung mit den Sekretären für Landwirtschaft in Bernburg," November 19, 1982.

93 Zatlin, *The Currency of Socialism*, 135.

94 SAPMO-BArch, DY 30/1533, p. 394. Food Production and Supply (1982). From the director of the VVB KLW, December 15,1982.

95 Dietrich, *Hammer, Zirkel, Gartenzaun*, 263–64.

96 SAPMO-BArch, DY 30/1533, pp. 138–40. Food Production and Supply (1982). "Handmaterial für Genossen Felfe zur Beratung mit den Sekretären für Landwirtschaft in Bernburg," November 19, 1982.

97 SAPMO-BArch, DY 30/1689, pp. 68–69. Monthly, Quarterly, and Annual Results for Livestock Production. Report, February 1, 1982.

98 SAPMO-BArch, DY 30/J IV 2/2/1931, p. 88. Protokolle des Politbüro des Zentralkomitees der Sozialistische Einheit Partei Deutschlands. Protokoll No. 5/1982, Sitzung des Politbüros, February 2, 1982, Reinschrift.

99 SAPMO-BArch, DY 30/J IV 2/2/1931, p. 88. Protokolle des Politbüro des Zentralkomitees der Sozialistische Einheit Partei Deutschlands. Protokoll No. 5/1982, Sitzung des Politbüros, February 2, 1982, Reinschrift.

100 SAPMO BArch DY, 30/J IV 2/2/1973, p. 31, in Dietrich, *Hammer, Zirkel, Gartenzaun*, 282.
101 SAPMO BArch DY 30/J IV 2/2/1973, p. 31, in Dietrich, *Hammer, Zirkel, Gartenzaun*, 282.
102 Orwell, *Animal Farm*.
103 Dale, *Between State Capitalism and Globalisation*, 226–27.
104 Judt, *Postwar*, 417.
105 Dale, *Between State Capitalism and Globalisation*, 222.
106 Dale, *Between State Capitalism and Globalisation*, 206–7.
107 Rathmer, *Dr. Alexander Schalck-Golodkowski*, 154.
108 Dale, *Between State Capitalism and Globalisation*, 204.
109 Dale, *Between State Capitalism and Globalisation*, 230.
110 Dale, *Between State Capitalism and Globalisation*, 229–30.
111 Orwell, *Animal Farm*, 180.

AFTERWORD: GARBAGE DUMP OF THE WEST

1 In German, the Stasi's informants were known as *Inoffizielle Mitarbeiter*.
2 Schönfelder, *Mit Gott Gegen Gülle*, 72. I draw upon Schönfelder's book on the protest movement in Neustadt/Orla to frame the introduction.
3 There is a voluminous body of literature on resistance and protests movements in the run-up to 1989. See Dale, *The East German Revolution of 1989*; Huff, *Natur und Industrie im Sozialismus;* Kallabis, *Ade, DDR!*; Neubert and Auerbach, *"Es kann anders werden"*; Ratzke, *Dagewesen und Aufgeschrieben*; Remy, *Opposition und Verweigerung in Nordthüringen*; and Maier, *Dissolution*.
4 Schönfelder, *Mit Gott gegen Gülle*, 51.
5 Maier, *Dissolution*, 61.
6 Schönfelder, *Mitt Gott gegen Gülle*, 133–34.
7 This is a story that has been told in many places. See Ash, *The Magic Lantern*; Fulbrook, *The People's State*; Hockenos, *Berlin* Calling; Jarausch, *The Rush to German Unity*; Judt, *Postwar*; and Maier, *Dissolution*.
8 Huff, *Nature und Industrie im Sozialismus*, 399. Huff argues that the end of the GDR traces back to the Stasi's anti-environmental actions in November 1987. See also Huff, "Environmental Policy in the GDR."
9 Uekötter, *The Greenest Nation?*, 127–28.
10 Uekötter, *The Greenest Nation?*
11 Jones, "Origins of the East German Environmental Movement"; Uekötter, *The Greenest Nation?*, 120–24.
12 Jones, "Origins," 248; Stoltzfus, "Public Space," 385–403.
13 Milder, *Greening Democracy*.
14 Uekötter, *The Greenest Nation?*, 114.
15 DeBardeleben, "'The Future Has Already Begun,'" 147.
16 Huff, *Natur und Industrie im Sozialismus*, 149, 357–58. Huff describes how Wilhelm Knabe, a scientist born in the GDR but who left for the FRG in 1959,

brought the issue of *Waldsterben* to national attention in the early 1980s and then, as a member of the Bundestag, smuggled environmental data to East German activists. For an international perspective on the rise, recognition, and combating of acid rain, see Rothschild, *Poisonous Skies*.

17 Jarausch, *The Rush to German Unity*, 37; Uekötter, *The Greenest Nation?*, 136.
18 Boehmer-Christiansen, "Environment-Friendly Deindustrialization," 78.
19 Petschow, Meyerhoff, Thomasberger, Borner, and Institut für Ökologische Wirtschaftsforschung, 9.
20 Petschow, Meyerhoff, Thomasberger, Borner, and Institut für Ökologische Wirtschaftsforschung, *Umweltreport DDR*, 11.
21 Helmut Kohl, "Fernsehansprache von Bundeskanzler Kohl Anläßlich des Inkrattretens der Währungs-, Wirtschafts- und Sozialunion," *Bulletin der Presse/ und Infomationsamt der Bundesregierung* 86 (July 3, 1990), www.helmut-kohl.de/index.php?msg=555. Film footage of a similar speech can be viewed at www.youtube.com/watch?v=Zhxeg-1ATfg.
22 Baylis, "Institutional Destruction and Reconstruction in Eastern Germany," 109–10.
23 Jarausch, *The Rush to German* Unity, 4–5.
24 Baylis, Institutional Destruction and Reconstruction in Eastern Germany," 18–21.
25 Smith, "The Illusory Economic Miracle," 123.
26 Peter Stegemann, "Keine Bank gibt den Bauern Kredit," *Die Tageszeitung*, May 22, 1990.
27 Buechler and Buechler, *Contesting Agriculture*, 178–79.
28 Smith, "The Illusory Economic Miracle," 131.
29 Dominick, "Capitalism, Communism, and Environmental Protection."
30 Morris, *Energy Democracy*, 131–32; Jarausch, *The Rush to German Unity*, 155.
31 Hochschule für nachhaltige Entwicklung, www.hnee.de/de/Fachbereiche/Fachbereiche-K258.htm.
32 Sundermann, Dorn, and Benning, "Wachsen oder trinken?" 202.
33 Wenz and Ziebarth, "Düngerüberschüsse aus der Landwirtschaft," 199–200.
34 Wenz and Ziebarth, "Düngerüberschüsse aus der Landwirtschaft," 199–200.
35 "Wildgewordener Keiler greift Fahrradfahrer an," *Frankfurter Neue Press*, March 11, 2013, www.fnp.de/rhein-main/Wildgewordener-Keiler-greift-Fahrradfahrer-an;art801,478702; Charles Hawley, "Radioactive Boar on the Rise in Germany," *Der Spiegel/ONLINE*, July 30, 2010, www.spiegel.de/international/zeitgeist/a-quarter-century-after-chernobyl-radioactive-boar-on-the-rise-in-germany-a-709345.html; Rainer W. During, "Wildschweinplage auf dem Friedhof dauert an: Erneut Gräber verwüstet," *Der Tagespiegel*, October 4, 2016, www.tagesspiegel.de/berlin/bezirke/spandau/wildschweinplage-auf-dem-friedhof-dauert-an-erneut-graeber-verwuestet/14640798.html; Kenny Langer and Annechristin Bonß, "Hund von Wildschwein angegriffen," *Sächsische Zeitung*, February 16, 2017, www.sz-online.de/nachrichten/hund-von-wildschweinen-angegriffen-3614162.html.

36 Juergen Batz, "Germany's Radioactive Boars a Legacy of Chernobyl," *Associated Press*, April 1, 2011, https://phys.org/news/2011-04-germany-radioactive-boars-legacy-chernobyl.html.
37 Boehmer-Christiansen, "Environment-Friendly Deindustrialization," 84.
38 Mellgrand, "Life after Lignite."
39 Fraser, *Rewilding the World*, 79–81.
40 Geidezis and Kreutz, "Green Belt Europe."
41 Annette Kuhn, "Berlin Zieht Wegen Schweinen vors Verfassungsgericht," September 26, 2017, www.morgenpost.de/berlin/article212057153/Berlin-zieht-wegen-Schweinen-vors-Verfassungsgericht.html.
42 Meyer-Renschhausen, *Urban Gardening in Berlin*, 10.
43 Sandra Schäfer and Nina Gessner, "Paradies Kleingarten: Die Neue Lust auf ein Leben als Laubenpieper," MOPO.de, August 13, 2017, www.mopo.de/hamburg/paradies-kleingarten-die-neue-lust-auf-ein-leben-als-laubenpieper-28122202.
44 Elena Krüskemper, "Integration am Gartenzaun: Deutsche Kleingärten Werden International," www.alumniportal-deutschland.org/deutschland/land-leute/kleingaerten, accessed September 5, 2018. This is also an issue I learned about from the research of my student Daria Lynch, who wrote and successfully defended an undergraduate honors thesis at the University of Rochester in 2019 titled "Gardeners into Germans: The Cultural Politics of the Schrebergarten, 1864–1939."
45 Meyer-Renschhausen and Holl, *Die Wiederkehr der Gärten*; Meyer-Renschhausen, Müller, Becker, and Arbeitsgruppe Kleins, *Die Gärten der Frauen*.
46 "The Pig Idea," accessed September 5, 2018, http://thepigidea.org.
47 Edwards and Driscoll, "From Farms to Factories."
48 Kendra Pierre-Louis, "Lagoons of Pig Waste Are Overflowing after Florence. Yes, That's as Nasty as It Sounds," *New York Times*, September 19, 2018, www.nytimes.com/2018/09/19/climate/florence-hog-farms.html.
49 Benjamin, *On the Concept of History*.

BIBLIOGRAPHY

ARCHIVES

Bundesarchiv Berlin (BArch)

DC 20	Council of Ministers
DE 2	State Administration for Statistics
DK 1	Ministry of Agriculture, Forestry, and Food Production
DK 5	Ministry for Environmental and Water Protection
DK 107	Academy of Agricultural Science
DK 108	Institute for Meat Production
DL 1	Ministry for Trade and Supply
DL 102	Institute for Market Research, Leipzig
KART 1004	General Map Collection of the GDR

Stiftung Archiv der Parteien und Massenorganisationen in Bundesarchiv (SAPMO-BArch)

DY 14	Union of Small Gardeners, Settlers, and Small Animal Breeders
DY 30	Socialist Unity Party of Germany
	Department of Agriculture
	Department of Trade, Supply, and Foreign Trade
	Office of Gerhard Grüneberg
	Office of Gunter Mittag
	Office of Werner Felfe
	Meetings of the Politburo and Central Committee
	Party Conferences of the SED

Landesarchiv Berlin (LAB)

C Rep. 112	Municipal Authority of Berlin, Department of Agriculture
C Rep. 117	Municipal Authority of Berlin, Department of Worker Provisioning and Food
C Rep. 147-01	Council of the City District Lichtenberg, Regional Mayor
C Rep. 902	Regional Chair of the SED, Berlin
C Rep. 903-01-03	District Leadership of the SED, Lichtenberg
C Rep. 904-062	Basic Organization of the SED, Press Office of the Government for the Chair of the Council of Ministers of the SED
C Rep. 904-251	Basic Organization of the SED, VVB Refrigerated and Warehouse Storage
C Rep. 904-289	Basic Organization of the SED, HO "Goods of Daily Need," Köpenick
C Rep. 904-311	Basic Organization of the SED, Economic Collective of Fruits, Vegetables, and Cooking Potatoes
C Rep. 950	Association of Mutual Farmer's Aid, Regional Association of Berlin

OTHER SOURCES

Adelman, Jeremy. "International Finance and Political Legitimacy: A Latin American View of the Global Shock." In *The Shock of the Global: The 1970s in Perspective*, edited by Niall Ferguson, 113–27. Cambridge, MA: Harvard University Press, 2010.

Ahrens, Ralf. "East German Foreign Trade in the Honecker Years." In *The East German Economy, 1945–2010: Falling Behind or Catching Up?*, edited by Hartmut Berghoff and Uta A. Balbier, 161–76. Cambridge, UK: Cambridge University Press, 2014.

Anderson, J. L. *Capitalist Pigs: Pigs, Pork, and Power in America*. Morgantown: West Virginia University Press, 2019.

———. *Industrializing the Corn Belt: Agriculture, Technology, and Environment, 1945–1972*. DeKalb: Northern Illinois University Press, 2009.

———. "Lard to Lean: Making the Meat-Type Hog in Post-World War II America." In *Food Chains: From Farmyard to Shopping Cart*, edited by Warren James Belasco and Roger Horowitz, 29-46. Philadelphia: University of Pennsylvania Press, 2009.

Anderson, Virginia DeJohn. *Creatures of Empire: How Domestic Animals Transformed Early America*. 1st ed. New York: Oxford University Press, 2006.

Applebaum, Anne. *Red Famine: Stalin's War on Ukraine*. New York: Doubleday, 2017.

Ash, Timothy Garton. *The Magic Lantern: The Revolution of '89 Witnessed in Warsaw, Budapest, Berlin, and Prague*. New York: Vintage, 1993.

Augustine, Dolores L. *Taking on Technocracy: Nuclear Power in Germany, 1945 to the Present*. New York: Berghahn Books, 2018.

Bange, Oliver, and Gottfried Niedhart. *Helsinki 1975 and the Transformation of Europe*. New York: Berghahn Books, 2008.

Bauerkämper, Arnd. "Abweichendes Verhalten in der Diktatur: Probleme einer Kategorialen Einordnung am Beispiel der Kollektivierung der Landwirtschaft in der DDR." In *Doppelte Zeitgeschichte: Deutsch-Deutsche Beziehungen 1945–1990*, edited by Arnd Bauerkämper, Martin Sabrow, and Bernd Stöver, 249–311. Bonn: J. H. W. Dietz, 1998.

———. "Amerikanisierung und Sowjetisierung in der Landwirtschaft. Zum Einfluß der Hegemonialmächte auf die Deutsche Agrarpolitik von 1945 bis zu den Frühen Sechziger Jahren." In *Amerikanisierung und Sowjetisierung in Deutschland 1945–1970*, edited by Konrad Hugo Jarausch and Hannes Siegrist, 195–218. New York: Campus, 1997.

Baylis, Thomas A. "Institutional Destruction and Reconstruction in Eastern Germany." In *After the Wall: Eastern Germany since 1989*, edited by Patricia Jo Smith, chapter 1 (n.p.). Boulder, CO: Westview Press, 1998.

Beaumont, Jaques. *Die DDR: Eine Chronik der Deutscher Geschichte*. Berlin: Otus Verlag: 2003.

Belasco, Warren James, and Roger Horowitz. *Food Chains: From Farmyard to Shopping Cart*. Philadelphia: University of Pennsylvania Press, 2009.

Benjamin, Walter. *On the Concept of History*. Createspace Independent Publishing Platform, 2016.

Berghoff, Hartmut, and Uta Andrea Balbier. *The East German Economy, 1945–2010: Falling Behind or Catching Up?* Cambridge, UK: Cambridge University Press, 2013.

Berry, Wendell. *The Unsettling of America: Culture & Agriculture*. San Francisco: Sierra Club Books, 1986.

Bess, Michael. *The Light-Green Society: Ecology and Technological Modernity in France, 1960–2000*. Chicago: University of Chicago Press, 2003.

Betts, Paul. *The Authority of Everyday Objects: A Cultural History of West German Industrial Design*. Berkeley: University of California Press, 2004.

Beyme, Klaus von. "Reconstruction in the German Democratic Republic." In *Rebuilding Europe's Bombed Cities*, edited by Jeffrey Diefendorf, 190–208. New York: Palgrave, 1990.

Bierber, Claudia, and Thomas Ruf. "Population Dynamics in Wild Boars *Sus Scrofa*: Ecology, Elasticity of Growth Rate and the Implications for the Management of Pulse Resource." *Journal of Applied Ecology* 42 (December 2005): 1203–13.

Bivar, Venus. *Organic Resistance: The Struggle over Industrial Farming in Postwar France*. Chapel Hill: University of North Carolina Press, 2018.

Blackbourn, David. *The Conquest of Nature: Water, Landscape, and the Making of Modern Germany*. New York: Norton, 2006.

Blanchette, Alex. "Herding Species: Biosecurity, Posthuman Labor, and the American Industrial Pig." *Cultural Anthropology* 30 (2015): 640–69.

Boehmer-Christiansen, Sonja. "Environment-Friendly Deindustrialization: Impacts of Unification on East Germany." In *Environment and Society in Eastern Europe*, edited by Andrew Tickle and Ian Welsh, 67–81. Boston: Addison Wesley, 1998.

Bornstein, Morris. *Comparative Economic Systems: Models and Cases*. 5th ed. Homewood, IL: Irwin, 1985.
Bramwell, Anna. *Blood and Soil: Richard Walther Darré and Hitler's "Green Party."* Buckinghamshire: Kensal Press, 1985.
Braudel, Fernand. *The Mediterranean and the Mediterranean World in the Age of Philip II*. Berkeley: University of California Press, 1995.
Bren, Paulina. *The Greengrocer and His TV: The Culture of Communism after the 1968 Prague Spring*. Ithaca, NY: Cornell University Press, 2010.
Bren, Paulina, and Mary Neuburger, eds. *Communism Unwrapped: Consumption in Cold War Eastern Europe*. New York: Oxford University Press, 2012.
Briedermann, Lutz. *Der Wildbestand, die Grosse Unbekannte: Methoden der Wildbestandsermittlung*. 1st ed. Stuttgart: Enke, 1983.
———. *Schwarzwild*. 1st ed. Berlin: Dt. Landwirtschaftsverl, 1985.
Brown, Frederick L. *The City Is More Than Human: An Animal History of Seattle*. Seattle: University of Washington Press, 2017.
Brown, Kate. "Gridded Lives: Why Kazakhstan and Montana Are Nearly the Same Place." *American Historical Review* 106 (2001): 17–48.
———. *Plutopia : Nuclear Families, Atomic Cities, and the Great Soviet and American Plutonium Disasters*. Oxford: Oxford University Press, 2013.
Brunner, Otto, Werner Conze, and Reinhart Koseleck. *Geschichtliche Grundbegriffe; Historisches Lexikon zur Politisch-Sozialen Sprache in Deutschland*. Stuttgart: E. Klett, 1972.
Buechler, Hans C., and Judith-Maria Buechler. *Contesting Agriculture: Cooperativism and Privatization in the New Eastern Germany*. Albany: State University of New York Press, 2002.
Bulliet, Richard W. *The Camel and the Wheel*. New York: Columbia University Press, 1975.
Burawoy, Michael, and János Lukács. *The Radiant Past: Ideology and Reality in Hungary's Road to Capitalism*. Chicago: University of Chicago Press, 1992.
Caldwell, Melissa L. *Dacha Idylls: Living Organically in Russia's Countryside*. Berkeley: University of California Press, 2010.
Caldwell, Peter C. *Dictatorship, State Planning, and Social Theory in the German Democratic Republic*. New York: Cambridge University Press, 2003.
Campbell, James B. *Introduction to Remote Sensing*. 4th ed. New York: Guildford Press, 2008.
Carlson, Peter. *K Blows Top: A Cold War Comic Interlude Starring Nikita Khrushchev, America's Most Unlikely Tourist*. 1st ed. New York: PublicAffairs, 2009.
Chaney, Sandra. *Nature of the Miracle Years: Conservation in West Germany, 1945–1975*. New York: Berghahn Books, 2008.
Ciesla, Burghard, and Helmut Suter. *Jagd und Macht: Die Geschichte des Jagdreviers Schorfheide*. Berlin: Be.Brad Verlag, 2011.
Confino, Alon. *The Nation as a Local Metaphor: Wurttemberg, Imperial Germany, and National Memory, 1871–1918*. Chapel Hill: University of North Carolina Press, 2000.

Corey, Lewis. *Meat and Man: A Study of Monopoly, Unionism, and Food Policy*. New York: Viking Press, 1950.

Cronon, William. *Changes in the Land: Indians, Colonists, and the Ecology of New England*. 1st ed. New York: Hill and Wang, 1983.

———. *Nature's Metropolis: Chicago and the Great West*. 1st ed. New York: Norton, 1991.

———. *Uncommon Ground : Toward Reinventing Nature*. 1st ed. New York: Norton, 1995.

Crosby, Alfred W. *Ecological Imperialism: The Biological Expansion of Europe, 900–1900*. New York: Cambridge University Press, 1986.

Cullather, Nick. *The Hungry World: America's Cold War Battle against Poverty in Asia*. Cambridge, MA: Harvard University Press, 2010.

Curry, Helen Anne. "Atoms in Agriculture: A Study of Scientific Innovation between Technological Systems." *Historical Studies in the Natural Sciences* 46 (January 1, 2015).

Dale, Gareth. *Between State Capitalism and Globalisation: The Collapse of the East German Economy*. New York: Peter Lang, 2004.

———. *The East German Revolution of 1989*. Manchester University Press, 2006.

Davidson, Osha Gray. *Broken Heartland: The Rise of America's Rural Ghetto*. Iowa City: University of Iowa Press, 1996.

DeBardeleben, Jean. "'The Future Has Already Begun:' Environmental Damage and Protection in the GDR." In *The Quality of Life in the German Democratic Republic: Changes and Developments in a State Socialist Society*, edited by Marilyn Rueschemeyer and Christiane Lemke, 144–64. Armonk, NY: M. E. Sharpe, 1989.

Dennis, Mike, and Eva Kolinsky. *United and Divided: Germany since 1990*. New York: Berghahn Books, 2004.

Dietrich, Isolde. *Hammer, Zirkel, Gartenzaun: Die Politik der SED Gegenüber den Kleingärten*. 1st ed. Berlin: Books on Demand, 2003.

Dominick, Raymond. "Capitalism, Communism, and Environmental Protection: Lessons from the German Experience." *Environmental History* 3 (July 1998): 311–32.

Edwards, Bob, and Adam Driscoll. "From Farms to Factories: The Environmental Consequences of Swine Industrialization in North Carolina." In *Twenty Lessons in Environmental Sociology*, edited by Kenneth Alan Gould and Tammy L. Lewis, 153–75. Oxford: Oxford University Press, 2009.

Ensminger, M. Eugene. *Swine Science*. 2nd ed. Danville, IL: Interstate, 1957.

———. *Swine Science*. 3d ed. Danville, IL: Interstate, 1961.

Essig, Mark. *Lesser Beasts: A Snout-to-Tail History of the Humble Pig*. New York: Basic Books, 2015.

Fabre-Vassas, Claudine. *The Singular Beast: Jews, Christians and the Pig*. New York: Columbia University Press, 1997.

Fairlie, Simon. *Meat: A Benign Extravagance*. White River Junction, VT: Chelsea Green, 2010.

Ferguson, Niall. *The Shock of the Global: The 1970s in Perspective*. Cambridge, MA: Belknap Press, 2011.

Ferrières, Madeleine. *Sacred Cow, Mad Cow: A History of Food Fears*. New York: Columbia University Press, 2006.

Fink, Hans-Georg, and Christoph Stubbe. *Die Jagd in der DDR: Ohne Pacht eine Andere Jagd*. Melsungen: Nimrod bei JANA, 2006.

Finlay, Mark. "The German Agricultural Experiment Stations and the Beginnings of American Agricultural Research." *Agricultural History* 62 (Spring 1988): 41–50.

———. "Hogs, Antibiotics, and the Industrial Environments of Postwar Agriculture." In *Industrializing Organisms: Introducing Evolutionary History*, edited by Philip Scranton and Susan R. Schrepfer, 237–60. New York: Routledge, 2004.

———. "New Sources, New Theses, and New Organizations in the New Germany: Recent Research on the History of German Agriculture." *Agricultural History* 75 (Summer 2001): 279–307.

Fitzgerald, Deborah K. "Eating and Remembering." *Agricultural History* 79 (Fall 2005): 393–408.

———. *Every Farm a Factory: The Industrial Ideal in American Agriculture*. New Haven: Yale University Press, 2003.

Fleischman, Thomas. "'A Plague of Wild Boars': A New History of Pigs and People in Late 20th Century Europe." *Antipode* 49, no. 4 (2017): 1015–34.

Foster, John Bellamy. *Marx's Ecology: Materialism and Nature*. New York: Monthly Review Press, 2000.

Francis, Richard C. *Domesticated: Evolution in a Man-Made World*. New York: Norton, 2015.

Frantz, Alaine C., Giovanna Massei, and Terry Burke. "Genetic Evidence for Past Hybridization between Domestic Pigs and English Wild Boars." *Conservation Genetics* 13 (October 2013): 1355–64.

Fraser, Caroline. *Rewilding the World: Dispatches from the Conservation Revolution*. 1st ed. New York: Metropolitan Books, 2009.

Frieden, Jeffry A. *Global Capitalism: Its Fall and Rise in the Twentieth Century*. 1st ed. New York: Norton, 2006.

Friedrich, Carl J., and Zbigniew Brzezinski. *Totalitarian Dictatorship and Autocracy*. 2nd ed. Cambridge, MA: Harvard University Press, 1965.

Fulbrook, Mary. *Anatomy of a Dictatorship: Inside the GDR, 1949–1989*. New York: Oxford University Press, 1995.

———. *The People's State: East German Society from Hitler to Honecker*. New Haven: Yale University Press, 2005.

Fulbrook, Mary, and Andrew I. Port. *Becoming East German: Socialist Structures and Sensibilities after Hitler*. New York: Berghahn Books, 2013.

Geidezis, Liana, and Melanie Kreutz. "Green Belt Europe—Nature Knows No Boundaries: From 'Iron Curtain' to Europe's Lifeline." *Urbani Izziv* 15, no. 2 (2004): 135–38.

Gibson, Abraham. *Feral Animals in the American South: An Evolutionary History*. Cambridge, UK: Cambridge University Press, 2016.

Gilbert, Jess. *Planning Democracy: Agrarian Intellectuals and the Intended New Deal.* New Haven: Yale University Press, 2015.
Gisolfi, Monica R. *The Takeover: Chicken Farming and the Roots of American Agribusiness.* Athens: University of Georgia Press, 2017.
Goldman, Mara, Paul Nadasdy, and Matt Turner. *Knowing Nature: Conversations at the Intersection of Political Ecology and Science Studies.* Chicago: University of Chicago Press, 2011.
Gopnik, Adam. "Honest, Decent, Wrong: The Invention of George Orwell." *New Yorker,* January 27, 2003. https://www.newyorker.com/magazine/2003/01/27/honest-decent-wrong.
Gorsuch, Anne E., and Diane Koenker, eds. *The Socialist Sixties: Crossing Borders in the Second World.* Bloomington: Indiana University Press, 2013.
Greene, Ann Norton. *Horses at Work: Harnessing Power in Industrial America.* Cambridge, MA: Harvard University Press, 2009.
Griffith, Nathan. *Husbandry: The Surest, Cheapest Way to Leisure, Plenty, Prosperity & Contentment.* Trout Valley, WV: Cobblemead Publications, 1998.
Groenen, Martien A. M. "A Decade of Pig Genome Sequencing: A Window on Pig Domestication and Evolution." *Genetics Selection Evolution* 48 (March 29, 2016): 23.
Groschoff, Kurt, and Richard Heinrich. *Die Landwirtschaft der DDR.* Berlin: Dietz, 1980.
Grove, Richard. *Green Imperialism: Colonial Expansion, Tropical Island Edens, and the Origins of Environmentalism, 1600–1860.* New York: Cambridge University Press, 1995.
Hamblin, Jacob. "Quickening Nature's Pulse: Atomic Agriculture at the International Atomic Energy Agency." *Dynamis* 35 (January 1, 2015): 389–408.
Hamilton, Shane. "Agribusiness, the Family Farm, and the Politics of Technological Determinism in the Post–World War II United States." *Technology and Culture* 55, no. 3 (August 8, 2014): 560–90.
———. "Supermarkets USA Confronts State Socialism: Airlifiting the Technopolitics of Industrial Food Distribution into Cold War Yugoslavia." In *Cold War Kitchen: Americanization, Technology, and European Users,* edited by Ruth Oldenziel and Karin Zachmann, 137–38. Cambridge, MA: MIT Press, 2009.
———. *Trucking Country: The Road to America's Wal-Mart Economy.* Politics and Society in Twentieth-Century America. Princeton: Princeton University Press, 2008.
Haney, David H. *When Modern Was Green: Life and Work of Landscape Architect Leberecht Migge.* New York: Routledge, 2010.
Haraway, Donna Jeanne. *Simians, Cyborgs, and Women: The Reinvention of Nature.* New York: Routledge, 1991.
———. *When Species Meet.* Minneapolis: University of Minnesota Press, 2008.
Harrison, Hope Millard. *Driving the Soviets up the Wall: Soviet-East German Relations, 1953–1961.* Princeton: Princeton University Press, 2003.
Harsch, Donna. *Revenge of the Domestic: Women, the Family, and Communism in the German Democratic Republic.* Princeton: Princeton University Press, 2007.

———. "Versorgungspolitik in der Sowjetisch Besetzten Zone." In *Hunger, Ernährung und Rationierungssysteme unter dem Staatssozialismus (1917–2006)*, edited by Matthias Middell and Felix Wemheuer, 213–43. Frankfurt am Main: Peter Lang, 2011.

Haselmann, Meike. "Die Jagd in der DDR—Zwischen Feudalismus und Sozialismus." In *VIII. Stipendiatkolloquium der Bundestiftung Aufarbeitung*, 39–43. Berlin, Germany, 2008. https://www.bundesstiftung-aufarbeitung.de/uploads/pdf-2008/reader_08.pdf.

Helleiner, Eric. *Forgotten Foundations of Bretton Woods: International Development and the Making of the Postwar Order.* Ithaca, NY: Cornell University Press, 2014.

Herf, Jeffrey. *Reactionary Modernism: Technology, Culture, and Politics in Weimar and the Third Reich.* New York: Cambridge University Press, 1984.

Hockenos, Paul. *Berlin Calling: A Story of Anarchy, Music, the Wall, and the Birth of the New Berlin.* New York: The New Press, 2017.

Holden, Palmer J., and M. Eugene Ensminger. *Swine Science.* 7th ed. Upper Saddle River, NJ: Pearson, 2005.

Hollsten, Laura. "Knowing Nature: Knowledge of Nature in Seventeenth Century French and English Travel Accounts from the Caribbean." PhD diss., Åbo Akademi, 2006.

Huff, Tobias. "Environmental Policy in the GDR: Principles, Restrictions, Failure, and Legacy." In *Ecologies of Socialisms: Germany, Nature, and the Left in History, Politics and Culture*, edited by Sabine Mödersheim, Scott Moranda, and Eli Rubin, 53–80. New York: Peter Lang, 2019.

———. *Natur und Industrie im Sozialismus: Eine Umweltgeschichte der DDR. Göttingen*, Germany: Vandenhoek & Ruprecht GmbH & Co., 2015.

Humphrey, Caroline. *The Unmaking of Soviet Life: Everyday Economies after Socialism.* Ithaca, NY: Cornell University Press, 2002.

Hurt, R. Douglas. *Problems of Plenty: The American Farmer in the Twentieth Century.* Chicago: Ivan R. Dee, 2002.

Ingram, Mrill. "Fermentation, Rot, and Other Human-Microbial Performances." In *Knowing Nature: Conversations at the Intersection of Political Ecology and Science Studies*, edited by Mara Goldman, Paul Nadasdy, and Matt Turner, 99–112. Chicago: University of Chicago Press, 2011.

Jacobsen, Hans-Dieter. "The Foreign Trade and Payments of the GDR in a Changing World Economy." In *The East German Economy*, edited by Ian Jeffries and Manfred Melzer, 235–60. New York: Croom Helm, 1987.

Jarausch, Konrad Hugo. "Care and Coercion: The GDR as Welfare Dictatorship." In *Dictatorship as Experience: Towards a Socio-Cultural History of the GDR*, edited by Jarausch, Konrad Hugo, 47–69. New York: Berghahn Books, 1999.

———, ed. *Dictatorship as Experience: Towards a Socio-Cultural History of the GDR.* New York: Berghahn Books, 1999.

———. *The Rush to German Unity.* New York: Oxford University Press, 1994.

Jasny, Naum. *The Socialized Agriculture of the USSR: Plans and Performance.* Stanford: Stanford University Press, 1949.

Johannes, Egon. *Entwicklung, Funktionswandel und Bedeutung Städtischer Kleingärten*. Kiel: Im Selbstverlag des Geographischen Instituts der Universität Kiel, 1955.

Jones, Merrill E. "Origins of the East German Environmental Movement." *German Studies Review* 16, no. 2 (1993): 235–64.

Jones, Susan D. *Death in a Small Package: A Short History of Anthrax*. Baltimore: Johns Hopkins University Press, 2010.

Josephson, Paul R. *An Environmental History of Russia*. New York: Cambridge University Press, 2013.

———. *Would Trotsky Wear a Bluetooth?: Technological Utopianism under Socialism, 1917–1989*. Baltimore: Johns Hopkins University Press, 2010.

Josling, Timothy Edward, and T. Geoffrey Taylor. *Banana Wars: The Anatomy of a Trade Dispute*. Boston: CABI, 2003.

Judt, Tony. *Postwar: A History of Europe since 1945*. New York: Penguin, 2005.

Kallabis, Heinz. *Ade, DDR!* Berlin: Treptower Verlaghaus, 1990.

Kideckel, David A. *The Solitude of Collectivism: Romanian Villagers to the Revolution and Beyond*. Ithaca, NY: Cornell University Press, 1993.

Kocha, Jürgen, Hartmut. "Eine Durchherrschte Gesellschaft." In *Sozialgeschichte der DDR*, edited by Jürgen Kocka, Hartmut Zwahr, and Hartmut Kaelble, 547–53. Stuttgart: Klett-Cotta /J. G. Cotta'sche Buchhandlung Nachfolger, 1994.

Koehler, John O. *Stasi: The Untold Story of the East German Secret Police*. Boulder, CO: Westview Press, 1999.

Kopstein, Jeffrey. *The Politics of Economic Decline in East Germany, 1945–1989*. Chapel Hill: University of North Carolina Press, 1997.

Kotkin, Stephen. "The Kiss of Debt: The East Bloc Goes Borrowing." In *The Shock of the Global: The 1970s in Perspective*, edited by Niall Ferguson, Charles S. Maier, Erez Manela, and Daniel J. Sargent, 80–93. Cambridge, MA: Belknap Press, 2010.

Lampe, John R., Russell O. Prickett, and Ljubiša S. Adamović. *Yugoslav-American Economic Relations Since World War II*. Durham, NC: Duke University Press, 1990.

Lange, Oskar, F. M. Taylor, and Benjamin Evans Lippincott. *On the Economic Theory of Socialism*. New York: McGraw-Hill, 1965.

Larson, Greger. "Worldwide Phylogeography of Wild Boar Reveals Multiple Centers of Pig Domestication." *Science* 307 (March 11, 2005): 1618–21.

Larson, Greger, Keith Dobney, Umberto Albarella, Meiying Fang, Elizabeth Matisoo-Smith, Judith Robins, Stewart Lowden, et al. "Worldwide Phylogeography of Wild Boar Reveals Multiple Centers of Pig Domestication." *Science* 307, no. 5715 (March 11, 2005): 1618–21.

Last, George. *After the "Socialist Spring": Collectivisation and Economic Transformation in the GDR*. New York: Berghahn Books, 2009.

Latour, Bruno. *Pandora's Hope: Essays on the Reality of Science Studies*. Cambridge, MA: Harvard University Press, 1999.

———. *The Pasteurization of France*. Cambridge, MA: Harvard University Press, 1988.

Laue, Anett. *Das Sozialistische Tier: Auswirkungen der SED-Politik auf Gesellschaftliche Mensch-Tier-Verhältnisse in der DDR (1949–1989)*. Köln: Böhlau Verlag, 2017.

Lebow, Katherine. *Unfinished Utopia: Nowa Huta, Stalinism, and Polish Society, 1949–56*. Ithaca, NY: Cornell University Press, 2013.

Lekan, Thomas M. *Imagining the Nation in Nature: Landscape Preservation and German Identity, 1885–1945*. Cambridge, MA: Harvard University Press, 2004.

Leopold, Aldo. *A Sand County Almanac: With Essays on Conservation from Round River*. Oxford: Oxford University Press, 1949.

Lindenberger, Thomas. *Volkspolizei. Herrschaftspraxis und Öffentliche Ordnung im SED-Staat 1952–1968*. Köln: Böhlau Verlag, 2003.

Lozon, Michael. *The Sun Never Sets on Big Dutchman*. Edited by Kathy Early. Muskegon, MI: Dobb Printing, 2004.

Luttrell, Clifton B. "The Russian Wheat Deal—Hindsight vs. Foresight." *Federal Reserve Bank of St. Louis Review*, October 1973.

Lutze, Gerd. *Landschaft im Wandel: Der Nordosten Brandenburgs vom 17. Jahrhundert bis Heute. Entdeckungen Entlang der Märkischen Eiszeitstraße*. Eberswalde, Germany: Gesellschaft zur Erforschung und Förderung der Märkischen Eiszeitstrasse, 2003.

Macmillan, Don. *The Big Book of John Deere Tractors*. St. Paul, MN: Voyageur Press, 1999.

Maier, Charles S. "'Als Wär's ein Stück von Uns . . .': German Politics and Society Traverses Twenty Years of United Germany." In *From the Bonn to the Berlin Republic: Germany at the Twentieth Anniversary of Unification*, edited by Jeffrey J. Anderson and Eric Langenbacher, 32–49. New York: Berghahn Books, 2010.

———. *Dissolution: The Crisis of Communism and the End of East Germany*. Princeton: Princeton University Press, 1999.

Major, Patrick, and Jonathan Osmond. *The Workers' and Peasants' State: Communism and Society in East Germany under Ulbricht, 1945–71*. Manchester: Manchester University Press, 2002.

Malcolmson, Robert, and Stephanos Mastoris. *The English Pig: A History*. London: The Hambledon Press, 1998.

Mayer, John J., and I. Lehr Brisbin. *Wild Pigs in the United States: Their History, Comparative Morphology, and Current Status*. Athens: University of Georgia Press, 1991.

McElvoy, Anne. *The Saddled Cow: East Germany's Life and Legacy*. London: Faber, 1993.

McNeur, Catherine. *Taming Manhattan: Environmental Battles in the Antebellum City*. Cambridge, MA: Harvard University Press, 2014.

Melville, Elinor G. K. *A Plague of Sheep: Environmental Consequences of the Conquest of Mexico*. Cambridge, UK: Cambridge University Press, 1997.

Meng, X. J., D. S. Lindsay, and N. Sriranganathan. "Wild Boars as Sources for Infectious Diseases in Livestock and Humans." *Philosophical Transactions of the Royal Society B: Biological Sciences* 364, no. 1530 (September 27, 2009): 2697–2707.

Merkel, Konrad. "Agriculture." In *The East German Economy*, edited by Ian Jeffries and Manfred Melzer, 202–34. London: Croom Helm, 1987.

Meyer-Renschhausen, Elisabeth. *Urban Gardening in Berlin: Touren zu den Neuen Gärten der Stadt*. Berlin: be.bra, 2016.

Meyer-Renschhausen, Elisabeth, and Anne Holl. *Die Wiederkehr der Gärten: Kleinlandwirtschaft im Zeitalter der Globalisierung*. Innsbruck: Studien Verlag, 2000.

Meyer-Renschhausen, Elisabeth, Renate Müller, Petra Becker, and Arbeitsgruppe Kleinstwirtschaft. *Die Gärten der Frauen: Zur Sozialen Bedeutung von Kleinstlandwirtschaft in Stadt und Land Weltweit*. Herbolzheim: Centaurus, 2002.

Meynhardt, Heinz. *Schwarzwild Report: Vier Jahre unter Schweine*. Berlin: Neumann-Neudamm Verlag, 1982.

Migge, Leberecht, and David H. Haney. *Garden Culture of the Twentieth Century*. Washington, DC: Dumbarton Oaks Research Library and Collection, 2013.

Milder, Stephen. *Greening Democracy: The Anti-nuclear Movement and Political Environmentalism in West Germany and Beyond, 1968–1983*. Cambridge, UK: Cambridge University Press, 2017.

Mizelle, Brett. *Pig*. London: Reaktion Books, 2011.

Moeller, Robert G. *German Peasants and Agrarian Politics, 1914–1924: The Rhineland and Westphalia*. Chapel Hill: University of North Carolina Press, 1986.

Moore, Jason W. *Capitalism in the Web of Life: Ecology and the Accumulation of Capital*. 1st ed. New York: Verso, 2015.

Moranda, Scott. *The People's Own Landscape: Nature, Tourism, and Dictatorship in East Germany*. Ann Arbor: University of Michigan Press, 2014.

Morgan, Dan. *Merchants of Grain: The Power and Profits of the Five Giant Companies at the Center of the World's Food Supply*. New York: Viking Press, 1979.

Morris, Craig. *Energy Democracy: Germany's Energiewende to Renewables*. 1st ed. London: Macmillan, 2016.

Naimark, Norman M. *The Russians in Germany: A History of the Soviet Zone of Occupation, 1945–1949*. Cambridge, MA: Harvard University Press, 1995.

Nance, Susan. *The Historical Animal*. Syracuse, NY: Syracuse University Press, 2015.

Nelson, Arvid. *Cold War Ecology: Forests, Farms, and People in the East German Landscape, 1945–1989*. New Haven: Yale University Press, 2005.

Nelson, Daniel. *Farm and Factory: Workers in the Midwest, 1880–1990*. Bloomington, IN: Indiana University Press. 1995.

Neubert, Ehrhart, and Thomas Auerbach. *"Es Kann Anders Werden": Opposition und Widerstand in Thüringen 1945–1989*. Köln: Böhlau Verlag, 2005.

Nitzsche, Gunther, and Thomas Paulke. *100 Jahre Schweinezuchtverband Brandenburg, Chronik 1913-2013*. Rulsdorf, Germany: Förderverein Deutsches Schweinemuseum, 2013.

Nolan, Mary. *The Transatlantic Century: Europe and America, 1890–2010*. Cambridge, UK: Cambridge University Press, 2012.

———. *Visions of Modernity: American Business and the Modernization of Germany*. New York: Oxford University Press, 1994.

Nove, Alec. *An Economic History of the USSR, 1917–1991*. 3rd ed. New York: Penguin Books, 1992.

———. "Will Russia Ever Feed Itself." *New York Times Magazine*, February 1, 1976, 209, 218–19.

Nunan, Timothy. *Humanitarian Invasion: Global Development in Cold War Afghanistan*. Cambridge, UK: Cambridge University Press, 2016.

Nuti, Leopoldo. *The Crisis of Détente in Europe: From Helsinki to Gorbachev, 1975–1985*. London: Routledge, 2009.

Oldenziel, Ruth, and Karin Zachmann, eds. *Cold War Kitchen: Americanization, Technology, and European Users*. Cambridge, MA: MIT Press, 2009.

Olmstead, Alan L., and Paul Webb Rhode. *Creating Abundance: Biological Innovation and American Agricultural Development*. New York: Cambridge University Press, 2008.

Opp, Karl-Dieter, Peter Voss, and Christiane Gern. *Origins of a Spontaneous Revolution: East Germany, 1989*. Ann Arbor: University of Michigan Press, 1995.

Orwell, George. *A Life in Letters*. Edited by Peter Hobley Davison. 1st American ed. New York: Liveright, 2013.

———. *Animal Farm*. New York: Harcourt, 1954.

Orwell, Sonia, and Ian Angus. *The Collected Essays, Journalism and Letters of George Orwell*. 4 vols. London: Secker & Warburg, 1968.

Osmond, Jonathan. "From Junker Estate to Co-operative Farm: East German Agrarian Society, 1945–1961." In *The Workers' and Peasants' State: Communism and Society in East Germany Under Ulbricht, 1945–71*, edited by Patrick Major and Jonathan Osmond, 130–50. Manchester: Manchester University Press, 2002.

Pachirat, Timothy. *Every Twelve Seconds: Industrialized Slaughter and the Politics of Sight*. New Haven: Yale University Press, 2011.

Patel, Kiran Klaus. *The New Deal: A Global History*. Princeton: Princeton University Press, 2017.

Patel, Raj. "The Long Green Revolution." *Journal of Peasant Studies* 40, no. 1 (January 2013): 1–63.

Patel, Raj, and Jason W. Moore. *A History of the World in Seven Cheap Things: A Guide to Capitalism, Nature, and the Future of the Planet*. Oakland: University of California Press, 2017.

Pence, Katherine, and Paul Betts. *Socialist Modern: East German Everyday Culture and Politics*. Ann Arbor: University of Michigan Press, 2008.

Petschow, Ulrich, Jürgen Meyerhoff, Klaus Thomasberger, Joachim Borner, and Institut für Ökologische Wirtschaftsforschung. *Umweltreport DDR: Bilanz der Zerstörung, Kosten der Sanierung, Strategien für den Ökologischen Umbau: Eine Studie des Instituts für Ökologische Wirtschaftsforschung*. Frankfurt am Main: S. Fischer, 1990.

Pollan, Michael. *The Omnivore's Dilemma: A Natural History of Four Meals*. New York: Penguin, 2006.

Port, Andrew I. *Conflict and Stability in the German Democratic Republic*. New York: Cambridge University Press, 2007.

Porter, Valerie. *Pigs: A Handbook to the Breeds of the World*. Ithaca, NY: Cornell University, 1993.
Poutrus, Patrice G. *Die Erfindung des Goldbroilers: Über den Zusammenhang Zwischen Herrschaftssicherung und Konsumentwicklung in der DDR*. Köln: Böhlau, 2002.
Premat, Adriana. *Sowing Change: The Making of Havana's Urban Agriculture*. Nashville, TN: Vanderbilt University Press, 2012.
Rathmer, Matthias. *Dr. Alexander Schalck-Golodkowski: Pragmatiker Zwischen den Fronten; eine Politische Biographie*. Berlin: epubli, 2015.
Ratzke, Dietrich. *Dagewesen und Aufgeschrieben: Reportagen über eine Deutsche Revolution*. IMK in der Verlag-Gruppe Frankfurter Allgemeine Zeitung, 1990.
Remy, Dietmar. *Opposition und Verweigerung in Nordthüringen (1976–1989)*. Dunderstadt: Mecke, 1999.
Ritvo, Harriet. *The Animal Estate: The English and Other Creatures in the Victorian Age*. Cambridge, MA: Harvard University Press, 1987.
Ritze, Werner. *Schweine: Zucht, Haltung, Fütterung*. Berlin: VEB Deutscher Landwirtschaftsverlag, 1971.
Rodden, John. *Understanding Animal Farm: A Student Casebook to Issues, Sources, and Historical Documents*. Westport, CT: Greenwood, 1999.
Rosenberg, Gabriel N. *The 4-H Harvest: Sexuality and the State in Rural America*. Philadelphia: University of Pennsylvania Press, 2015.
———. "A Race Suicide among the Hogs: The Biopolitics of Pork in the United States, 1865–1930." *American Quarterly* 68, no. 1: 49–73.
Ross, Corey. *Constructing Socialism at the Grass-Roots: The Transformation of East Germany, 1945-65*. New York: St. Martin's, 2000.
Rothschild, Rachel Emma. *Poisonous Skies: Acid Rain and the Globalization of Pollution*. Chicago: University of Chicago Press, 2019.
Rubin, Eli. *Synthetic Socialism: Plastics & Dictatorship in the German Democratic Republic*. Chapel Hill: University of North Carolina Press, 2008.
———. "Understanding a Car in the Context of a System: Trabants, Marzahn, and East German Socialism." In *The Socialist Car: Automobility in the Eastern Bloc*, edited by Lewis H. Siegelbaum, 124–41. Ithaca, NY: Cornell University Press, 2013.
Sachse, Carola. "Bullen, Hengste, Wissenschaftler. Diplomatische Tiere im Kalten Krieg." In *Wandlungen und Brüche: Wissenschaftsgeschichte als Politische Geschichte*, edited by Johannes Feichtinger, Marianne Klemun, Jan Surman, and Petra Svatek, 345–54. Göttingen: Vandenhoeck & Ruprecht, 2018.
Saraiva, Tiago. *Fascist Pigs: Technoscientific Organisms and the History of Fascism*. Cambridge, MA: MIT Press, 2016.
Saunders, Frances Stonor. *The Cultural Cold War: The CIA and the World of Arts and Letters*. New York: The New Press, 2013.
Schabel, Hans G. "Deer and Dauerwald in Germany: Any Progress?" *Wildlife Society Bulletin* 13 (Autumn 2001): 888–98.
Schier, Barbara. *Alltagsleben im "Sozialistischen Dorf": Merxleben und Seine LPG im Spannungsfeld der SED-Agrarpolitik, 1945–1990*. Berlin: Waxmann Verlag, 2001.

Schmalzer, Sigrid. *Red Revolution, Green Revolution: Scientific Farming in Socialist China*. Chicago: University of Chicago Press, 2016.

Schmidt, Helmut, and Fritz Richard Stern. *Unser Jahrhundert: Ein Gespräch*. München: Beck, 2010.

Schneider, Peter. *The Wall Jumper: A Berlin Story*. Chicago: University of Chicago Press, 1998.

Schönfelder, Jan. *Mit Gott Gegen Gülle: Die Umweltgruppe Knau/Dittersdorf 1986 bis 1991: Eine Regionale Protestbewegung in der DDR*. Vol. 7 of *Beiträge zur Geschichte und Stadtkultur*. Rudolstadt: Hain, 2000.

Schürer, Gerhard. *Gewagt und Verloren: Eine Deutsche Biographie*. Frankfurt/Oder: Frankfurter Oder Editionen, 1996.

Schwartz, Thomas Alan. "Legacies of Detente: A Three-Way Discussion." *Cold War History* 8 (November 2008): 513–25.

Schwarzer, Oskar. *Sozialistische Zentralplanwirtschaft in der SBZ/DDR: Ergebnisse Eines Ordnungspolitischen Experiments (1945–1989)*. Stuttgart: F. Steiner, 1999.

Scott, James C. *Seeing like a State: How Certain Schemes to Improve the Human Condition Have Failed*. New Haven: Yale University Press, 1998.

Scranton, Philip, and Susan R. Schrepfer. *Industrializing Organisms: Introducing Evolutionary History*. New York: Routledge, 2004.

Shrubb, Michael. *Birds, Scythes and Combines: A History of Birds and Agricultural Change*. Cambridge, UK: Cambridge University Press, 2003.

Siegelbaum, Lewis H. *The Socialist Car: Automobility in the Eastern Bloc*. Ithaca, NY: Cornell University Press, 2013.

Siegemund, Klaus, *Die Individuelle Produktion Landwirtschaftlicher Erzeugnisse in der DDR und Vorschläge zur Vervollkommung Ihrer Ökonomische Stimulierung*. PhD diss., university unknown, November, 1987.

Slaveski, Filip. *The Soviet Occupation of Germany: Hunger, Mass Violence and the Struggle for Peace, 1945–1947*. Cambridge, UK: Cambridge University Press, 2013.

Smith, Jenny Leigh. *Works in Progress: Plans and Realities on Soviet Farms, 1930–1963*. New Haven: Yale University Press, 2014.

Smith, Patricia Jo. "The Illusory Economic Miracle: Assessing Eastern Germany's Economic Transition." In *After the Wall: Eastern Germany since 1989*, edited by Patricia Jo Smith, 109–39. Boulder, CO: Westview Press, 1998.

Soule, Judith D., and Jon K. Piper. *Farming in Nature's Image: An Ecological Approach to Agriculture*. Washington, DC: Island Press, 1992.

Specht, Joshua. *Red Meat Republic: A Hoof-to-Table History of How Beef Changed America*. Princeton: Princeton University Press, 2019.

Spufford, Francis. *Red Plenty*. London: Faber, 2010.

Steil, Benn. *The Battle of Bretton Woods: John Maynard Keynes, Harry Dexter White, and the Making of a New World Order*. Princeton: Princeton University Press, 2013.

Stein, Judith. *Pivotal Decade: How the United States Traded Factories for Finance in the Seventies*. New Haven: Yale University Press, 2010.

Steiner, André. *The Plans That Failed: An Economic History of the GDR*. Vol. 13 of *Studies in German History*. New York: Berghahn Books, 2013.

Stitziel, Judd. *Fashioning Socialism: Clothing, Politics, and Consumer Culture in East Germany*. Oxford: Berg, 2005.

Stokes, Raymond G. *Constructing Socialism: Technology and Change in East Germany 1945–1990*. Baltimore: Johns Hopkins University Press, 2000.

———. "From Schadenfreude to Going-Out-of-Business Sale: East Germany and the Oil Crises of the 1970s." In *The East German Economy, 1945–2010: Falling Behind or Catching Up?*, edited by Hartmut Berghoff and Uta A. Balbier, 131–44. Cambridge, UK: Cambridge University Press, 2014.

Stoll, Steven. *The Fruits of Natural Advantage: Making the Industrial Countryside in California*. Berkeley: University of California Press, 1998.

———. *Larding the Lean Earth: Soil and Society in Nineteenth-Century America*. 1st ed. New York: Hill and Wang, 2002.

Stoltzfus, Nathan. "Public Space and the Dynamics of Environmental Action: Green Protest in the German Democratic Republic." *Archiv für Sozialgeschichte* 43 (March 2003): 385–403.

Stubbe, Christoph. "Ausbildung und Schulung." In *Die Jagd in der DDR: Ohne Pacht eine Andere Jagd*, edited by Hans-Georg Fink and Christoph Stubbe, 283–94. Melsungen: Nimrod bei JANA, 2006.

———. "Historischer Abriss der Organisation des Jagdwesens von 1946 bis 1962 in der Sowjetischen Besatzungszone und in der DDR." In *Die Jagd in der DDR: Ohne Pacht eine Andere Jagd*, edited by Hans-Georg Fink and Christoph Stubbe, 12–29. Melsungen: Nimrod bei JANA, 2006.

———. "Wildbewirtschaftung." In *Die Jagd in der DDR: Ohne Pacht Eine Andere Jagd*, edited by Hans-Georg Fink and Christoph Stubbe, 69–143. Melsungen: Nimrod bei JANA, 2006.

Sundermann, Jutta, Leonie Dorn, and Reinheld Benning. "Wachsen oder Trinken? Grenzen der Konzentration in die Tierhaltung und der Streit um die Düngeverordnung." *Der Kritische Agrarbericht 2016* 15 (2016): 200–5.

Suri, Jeremi. *Power and Protest: Global Revolution and the Rise of Detente*. Cambridge, MA: Harvard University Press, 2003.

Taiganides, E. Paul, *Pig Waste Management and Recycling: The Singapore Experience*, Ottawa: International Development Research Centre, 1992.

Taylor, Fred. *Exorcising Hitler: The Occupation and Denazification of Germany*. London: Bloomsbury, 2011.

Taylor, Joseph E. *Making Salmon: An Environmental History of the Northwest Fisheries Crisis*. Seattle, WA: University of Washington Press, 1999.

Terent'eva, Aleksandra Semenevna. *Zu Einigen Fragen der Intensivierung der Industriemässigen Schweineproduktion*, Vol. 16, *Fortschrittsberichte für die Landwirtschaft und Nahrungsgüterwirtschaft 4*. Berlin: Akademie der Landwirtschaftswissenschaften der Deutschen Demokratischen Republik, 1978.

Tooze, J. Adam. *The Wages of Destruction: The Making and Breaking of the Nazi Economy*. New York: Allen Lane, 2006.

Trager, James. *Amber Waves of Grain*. New York: Arthur Fields Books, 1973.

Trömer, Bruno. "Gesetzgebung, Organisation und Leitung des Jagdwesens Ab 1962 (Bildung der Jagdgesellschaft)." In *Die Jagd in der DDR: Ohne Pacht eine Andere Jagd*, edited by Hans-Georg Fink and Christoph Stubbe, 30–68. Melsungen: Nimrod bei JANA, 2006.
Turner, Matthew D. "Introduction." In *Knowing Nature: Conversations at the Intersection of Political Ecology and Science Studies*, edited by Mara Goldman, Paul Nadasdy, and Matt Turner, 1–23. Chicago: University of Chicago Press, 2011.
Uekötter, Frank. *Die Wahrheit Ist auf dem Feld: Eine Wissensgeschichte der Deutschen Landwirtschaft*. Göttingen: Vandenhoeck & Ruprecht, 2011.
———. *The Green and the Brown: A History of Conservation in Nazi Germany*. New York: Cambridge University Press, 2006.
———. *The Greenest Nation? A New History of German Environmentalism*. Cambridge, MA: The MIT Press, 2014.
Unger, Bernd D. W. *Der Berliner Bär*. Münster: Waxmann Verlag, 2000.
Verdery, Katherine. *The Vanishing Hectare: Property and Value in Postsocialist Transylvania*. Ithaca, NY: Cornell University Press, 2003.
———. *What Was Socialism, and What Comes Next?* Princeton: Princeton University Press, 1996.
Virgos, Emilio. "Factors Affecting Wild Boar (*Sus Scrofa*) Occurrence in Highly Fragmented Mediterraenan Landscapes." *Canadian Journal of Zoology* 80 (March 2002): 430–35.
Walker, Brett L. "Commercial Growth and Environmental Change in Early Modern Japan: Hachinohe's Wild Boar Famine." *The Journal of Asian Studies* 60 (May 2001).
———. *The Lost Wolves of Japan*. Seattle: University of Washington Press, 2005.
Walker, Richard. *The Conquest of Bread: 150 Years of Agribusiness in California*. New York: New Press, 2004.
Watson, Lyall. *The Whole Hog: Exploring the Extraordinary Potential of Pigs*. Washington, DC: Smithsonian Books, 2004.
Weiher, Ottfried, and Christoph Langner. *Tierzucht in Voermpommern von 1945 bis zur II. Pommernschau 2012*. Friedland, Germany: Verein Pommernschau e.V., 2012.
Weis, Anthony John. *The Ecological Hoofprint: The Global Burden of Industrial Livestock*. London: Zed Books, 2013.
Wenz, Katrin, and Katrin Ziebarth. "Düngerüberschüsse aus der Landwirtschaft: Gefahr für Flüsse, Seen, und Meere." *Der Kritische Agrarbericht 2017* 16 (January 2017).
West, Brian, and Brian Kennedy. *Swine Manure Handling Systems*. Edmonton: Alberta Agriculture, Engineering Field Services Branch, 1983.
Whatmore, Sarah. *Hybrid Geographies: Natures, Cultures, Spaces*. Thousand Oaks, CA: Sage, 2002.
White, Richard. *The Organic Machine*. New York: Hill and Wang, 1995.
White, Sam. "From Globalized Pig Breeds to Capitalist Pigs: A Study in Animal Cultures and Evolutionary History." *Environmental History* 16 (January 2011): 94–120.

Willes, Margaret. *The Gardens of the British Working Class*. New Haven: Yale University Press, 2014.

Woods, Abigail. *A Manufactured Plague? The History of Foot and Mouth Disease in Britain*. Sterling, VA: Earthscan, 2004.

Woods, Rebecca J. H. *The Herds Shot Round the World: Native Breeds and the British Empire, 1800–1900*. Chapel Hill: University of North Carolina Press, 2017.

Worster, Donald. *Dust Bowl: The Southern Plains in the 1930s*. New York: Oxford University Press, 2004.

———. *Rivers of Empire: Water, Aridity, and the Growth of the American West*. New York: Oxford University Press, 1992.

Zachmann, Karin. *Mobilisierung der Frauen: Technik, Geschlecht und Kalter Krieg in der DDR*. Reihe "Geschichte und Geschlechter," vol. 44. Frankfurt/Main: Campus, 2004.

Zatlin, Jonathan R. *The Currency of Socialism: Money and Political Culture in East Germany*. Cambridge, UK: Cambridge University Press, 2007.

Zeisler-Vralsted, Dorothy. *Rivers, Memory, and Nation-Building: A History of the Volga and Mississippi Rivers*. Vol. 4 of *The Environment in History: International Perspectives*. New York: Berghahn Books, 2015.

Zimmerman, Andrew. *Alabama in Africa: Booker T. Washington, the German Empire, and the Globalization of the New South*. Princeton: Princeton University Press, 2010.

Zimmerman, Herbert. "Unraveling the Ties That Really Bind: The Dissolution of the Transatlantic Monetary Order and the European Monetary Cooperation." In *The Strained Alliance: U.S.-European Relations from Nixon to Carter*, edited by Matthias Schulz and Thomas Alan Schwartz, 125–44. Washington, DC: Cambridge University Press, 2010.

Zubok, Vladislav. "The Soviet Union and Detente of the 1970s." *Cold War History* 8 (2008): 427–47.

INDEX

WEYERHAEUSER ENVIRONMENTAL BOOKS

Seeds of Control: Japan's Empire of Forestry in Colonial Korea, by David Fedman

Fir and Empire: The Transformation of Forests in Early Modern China, by Ian M. Miller

Communist Pigs: An Animal History of East Germany's Rise and Fall, by Thomas Fleischman

Footprints of War: Militarized Landscapes in Vietnam, by David Biggs

Cultivating Nature: The Conservation of a Valencian Working Landscape, by Sarah R. Hamilton

Bringing Whales Ashore: Oceans and the Environment of Early Modern Japan, by Jakobina K. Arch

The Organic Profit: Rodale and the Making of Marketplace Environmentalism, by Andrew N. Case

Seismic City: An Environmental History of San Francisco's 1906 Earthquake, by Joanna L. Dyl

Smell Detectives: An Olfactory History of Nineteenth-Century Urban America, by Melanie A. Kiechle

Defending Giants: The Redwood Wars and the Transformation of American Environmental Politics, by Darren Frederick Speece

The City Is More Than Human: An Animal History of Seattle, by Frederick L. Brown

Wilderburbs: Communities on Nature's Edge, by Lincoln Bramwell

How to Read the American West: A Field Guide, by William Wyckoff

Behind the Curve: Science and the Politics of Global Warming, by Joshua P. Howe

Whales and Nations: Environmental Diplomacy on the High Seas, by Kurkpatrick Dorsey

Loving Nature, Fearing the State: Environmentalism and Antigovernment Politics before Reagan, by Brian Allen Drake

Pests in the City: Flies, Bedbugs, Cockroaches, and Rats, by Dawn Day Biehler

Tangled Roots: The Appalachian Trail and American Environmental Politics, by Sarah Mittlefehldt

Vacationland: Tourism and Environment in the Colorado High Country, by William Philpott

Car Country: An Environmental History, by Christopher W. Wells

Nature Next Door: Cities and Trees in the American Northeast, by Ellen Stroud

Pumpkin: The Curious History of an American Icon, by Cindy Ott

The Promise of Wilderness: American Environmental Politics since 1964, by James Morton Turner

The Republic of Nature: An Environmental History of the United States, by Mark Fiege

A Storied Wilderness: Rewilding the Apostle Islands, by James W. Feldman

Iceland Imagined: Nature, Culture, and Storytelling in the North Atlantic, by Karen Oslund

Quagmire: Nation-Building and Nature in the Mekong Delta, by David Biggs

Seeking Refuge: Birds and Landscapes of the Pacific Flyway, by Robert M. Wilson

Toxic Archipelago: A History of Industrial Disease in Japan, by Brett L. Walker

Dreaming of Sheep in Navajo Country, by Marsha L. Weisiger

Shaping the Shoreline: Fisheries and Tourism on the Monterey Coast, by Connie Y. Chiang

The Fishermen's Frontier: People and Salmon in Southeast Alaska, by David F. Arnold

Making Mountains: New York City and the Catskills, by David Stradling

Plowed Under: Agriculture and Environment in the Palouse, by Andrew P. Duffin

The Country in the City: The Greening of the San Francisco Bay Area, by Richard A. Walker

Native Seattle: Histories from the Crossing-Over Place, by Coll Thrush

Drawing Lines in the Forest: Creating Wilderness Areas in the Pacific Northwest, by Kevin R. Marsh

Public Power, Private Dams: The Hells Canyon High Dam Controversy, by Karl Boyd Brooks

Windshield Wilderness: Cars, Roads, and Nature in Washington's National Parks, by David Louter

On the Road Again: Montana's Changing Landscape, by William Wyckoff

Wilderness Forever: Howard Zahniser and the Path to the Wilderness Act, by Mark Harvey

The Lost Wolves of Japan, by Brett L. Walker

Landscapes of Conflict: The Oregon Story, 1940-2000, by William G. Robbins

Faith in Nature: Environmentalism as Religious Quest, by Thomas R. Dunlap

The Nature of Gold: An Environmental History of the Klondike Gold Rush, by Kathryn Morse

Where Land and Water Meet: A Western Landscape Transformed, by Nancy Langston

The Rhine: An Eco-Biography, 1815-2000, by Mark Cioc

Driven Wild: How the Fight against Automobiles Launched the Modern Wilderness Movement, by Paul S. Sutter

George Perkins Marsh: Prophet of Conservation, by David Lowenthal

Making Salmon: An Environmental History of the Northwest Fisheries Crisis, by Joseph E. Taylor III

Irrigated Eden: The Making of an Agricultural Landscape in the American West, by Mark Fiege

The Dawn of Conservation Diplomacy: U.S.-Canadian Wildlife Protection Treaties in the Progressive Era, by Kirkpatrick Dorsey

Landscapes of Promise: The Oregon Story, 1800-1940, by William G. Robbins

Forest Dreams, Forest Nightmares: The Paradox of Old Growth in the Inland West, by Nancy Langston

The Natural History of Puget Sound Country, by Arthur R. Kruckeberg

WEYERHAEUSER ENVIRONMENTAL CLASSICS

Environmental Justice in Postwar America: A Documentary Reader, edited by Christopher W. Wells

Making Climate Change History: Documents from Global Warming's Past, edited by Joshua P. Howe

Nuclear Reactions: Documenting American Encounters with Nuclear Energy, edited by James W. Feldman

The Wilderness Writings of Howard Zahniser, edited by Mark Harvey

The Environmental Moment: 1968-1972, edited by David Stradling

Reel Nature: America's Romance with Wildlife on Film, by Gregg Mitman

DDT, Silent Spring, and the Rise of Environmentalism, edited by Thomas R. Dunlap

Conservation in the Progressive Era: Classic Texts, edited by David Stradling

Man and Nature: Or, Physical Geography as Modified by Human Action, by George Perkins Marsh

A Symbol of Wilderness: Echo Park and the American Conservation Movement, by Mark W. T. Harvey

Tutira: The Story of a New Zealand Sheep Station, by Herbert Guthrie-Smith

Mountain Gloom and Mountain Glory: The Development of the Aesthetics of the Infinite, by Marjorie Hope Nicolson

The Great Columbia Plain: A Historical Geography, 1805-1910, by Donald W. Meinig

CYCLE OF FIRE

Fire: A Brief History, second edition, by Stephen J. Pyne

The Ice: A Journey to Antarctica, by Stephen J. Pyne

Burning Bush: A Fire History of Australia, by Stephen J. Pyne

Fire in America: A Cultural History of Wildland and Rural Fire, by Stephen J. Pyne

Vestal Fire: An Environmental History, Told through Fire, of Europe and Europe's Encounter with the World, by Stephen J. Pyne

World Fire: The Culture of Fire on Earth, by Stephen J. Pyne

ALSO AVAILABLE:

Awful Splendour: A Fire History of Canada, by Stephen J. Pyne